Managing Physical Plant Operations

MANAGING PHYSICAL PLANT OPERATIONS

BY KENNETH PETROCELLY, C.P.E., C.F.M.

Published by
THE FAIRMONT PRESS, INC.
700 Indian Trail
Lilburn, GA 30247

Library of Congress Cataloging-in-Publication Data

Petrocelly, K. L. (Kenneth Lee), 1946-
  Managing physical plant operations / by Kenneth Petrocelly.
  p. cm.
  Includes index.
  ISBN 0-88173-160-9
  1. Power-plants--Management. I. Title.
TJ164.P48   1993   658.2'6--dc20   93-35434
                                              CIP

Managing Physical Plant Operations by Kenneth Petrocelly.
©1994 by The Fairmont Press, Inc. All rights reserved. No part of this publication may be reproduced or transmitted in any form or by any means, electronic or mechanical, including photocopy, recording, or any information storage and retrieval system, without permission in writing from the publisher.

Published by The Fairmont Press, Inc.
700 Indian Trail
Lilburn, GA 30247

Printed in the United States of America

10 9 8 7 6 5 4 3 2 1

ISBN 0-88173-160-9 FP

ISBN 0-13-147455-3 PH

While every effort is made to provide dependable information, the publisher, authors, and editors cannot be held responsible for any errors or omissions.

Distributed by PTR Prentice Hall
Prentice-Hall, Inc.
A Paramount Communications Company
Englewood Cliffs, NJ 07632

Prentice-Hall International (UK) Limited, London
Prentice-Hall of Australia Pty. Limited, Sydney
Prentice-Hall Canada Inc., Toronto
Prentice-Hall Hispanoamericana, S.A., Mexico
Prentice-Hall of India Private Limited, New Delhi
Prentice-Hall of Japan, Inc., Tokyo
Simon & Schuster Asia Pte. Ltd., Singapore
Editora Prentice-Hall do Brasil, Ltda., Rio de Janeiro

To Susan,
who managed the "home" plant so well
while I was away.

Table of Contents

Chapter 1　　THE POWER PLANT
Welcome To My World, A Historical Perspective,
Touring The Facility, Some Initial Considerations,
Tentative Implementation Plan 1

Chapter 2　　STANDARD OPERATING PROCEDURES
Setting Up A System, Section Modularity, Typical Policy
Topics... *Department Operations, Financial/Contracts,
Personnel Related, In-Service/Training, Work Order
System, Equipment Management, Preventive Maintenance,
Safety Management, Fire Protection, Emergency
Preparedness, Service Agreements, Regulatory
Compliance, Energy Management, Committee
Membership, Quality Assurance* 15

Chapter 3　　CONTRACT SERVICES
Pertinent Focal Points, Proposals Versus Quotations,
Request For Proposal, Request For Quotation,
The Boiler Plate 27

Chapter 4　　COMPUTERIZED MAINTENANCE
The PM Function, Information Gathering, Choosing A
Software Package, Caveat Emptor, Installing
The Program 53

Chapter 5　　MANPOWER REQUIREMENTS
The Screening Process, Position Descriptions,
Licenses And Permits, Staffing Levels, Orientation
And Training 63

Chapter 6　　BOILER BASICS
Their Support Systems, Internal And External Inspections,
Sound Operating Principles, General Room
Requirements, Increasing Operating Efficiency 89

Chapter 7　　CHILLER ESSENTIALS
System Component Variations, Mechanical
Refrigeration Standards, Installation Specs, Assuring

Operating Integrity, Psychrometric Observations,
Cold Weather Operations 101

Chapter 8   AUXILIARY/SUPPORT EQUIPMENT
Compressor Accessories, Cooling Tower Structures,
Electric Motor Selection, Generator Maintenance,
Pump Repairs, Turbine Inspection 113

Chapter 9   ELECTROMECHANICAL ROOMS
Air Handling Units, Electric Power Systems,
Uninterruptible Power Supplies, Pressure Relief Valves,
Steam Trapping, Equipment/Room Checks 125

Chapter 10   LOSS PREVENTION/RECOVERY
Boiler and Machinery Insurance, Investigation of
Accidents, Causes of Accidents, Pressure Vessel
Failures, Electromechanical Failures, Boiler Control
Failures, Boiler Explosions 139

Chapter 11   EQUIPMENT CONSERVATION
Evaluating The Operation, Troubleshooting System
Design, Contingency Planning, The Concept of
Maintenance, Nondestructive Testing 151

Chapter 12   ENERGY CONSUMPTION
The Building Envelope, Equipment Operation and
Maintenance, General Conservation Measures, Construction
Considerations, Energy Management Systems 165

APPENDICES
Appendix A   RFQ Boiler Plate 175
Appendix B   Standard Preventive Maintenance
              Assignments 201
Appendix C   Special Operating Concerns 217
Appendix D   Stationary Engineer Duties & Traits 231
Appendix E   Universal Task Instructions 245
Appendix F   Tables 259

GLOSSARY 285
INDEX 307

Foreword

Of the eight books I've written on the subject, this wasn't the most painstaking of the bunch, but the project that spawned it was the most painful I've ever experienced. Considered neither a renovation nor new construction, the governing body responsible for its reopening let separate contracts to as many firms in an attempt to restore the physical plant of a failed health care institution. Individual contractors converged (*en masse*) on the property, without benefit of guidance or coordination of a general. My job? To make sure they had done theirs. My pain? I arrived 2 years into the project, had no "clerk of the works" to confer with, was not aware of agreed upon work scopes (nor were the contractors interested in educating me), had no blueprints to work with and no documented history to give me a handle on past performance (or non-performance for that matter). The good news? Somehow we all survived each other, received a certificate of occupancy, moved out the gang boxes and moved in the operating staff. The bad news? I've been remobilized to hold their hands during the licensing survey in June, official opening in July and the accreditation inspection in August. — Why do we do these kinds of things to ourselves?

CHAPTER 1

THE POWER PLANT

Caught up in tenant squabbles, suite renovations, employee grievances and budget constraints, Facilities Managers are often tempted to allow their power plants to run freely—under their own power so to speak. After all, everything down there is on autopilot anyway, isn't it? Well... not exactly. Make yourself comfortable: this won't be strenuous but it may take a while.

Welcome To My World

I'm in the midst of a major reconstruction project to bring a 10-year-old facility that's been closed down for the past five, back from the dead and into code compliance. And the timing couldn't have been better, for I got the nod to write this book just before I was hired on to oversee the job. So instead of creating scenarios to get my points across, as I have in my prior works (you have read them, haven't you?), I'll be relaying the essence of the information contained in the entries made in my daily journal, as the work progresses. As the focus of this book is on managing and operating the power plant, I'll do my best to limit our discussions to those issues, on that topic. If I should begin to stray into concerns I'm having regarding other aspects of the project... just bang on some ductwork with a ballpeen hammer; that'll rattle me and get me back on track. Shall we get started? Great!

A Historical Perspective

Before we begin the walk-through, let me introduce you to Allen Boldt, the Chief Engineer. Allen will field any questions regarding the history and present operation of the power house. He's been here for 2 years; originally hired on to stem the rapid deterioration of the plant's systems and to assure heat was provided within the structure during the winter months.

According to Allen, the facility is 10 years old. It was quickly constructed with a total disregard for then existing codes and ordinances but was somehow allowed to open its doors to the public by way of executive order. In its construction, the workmanship displayed was shoddy at best; only the cheapest materials were used and no inventory was ever established for repair work. In subsequent years, no money was ever reinjected into the operation and its systems simply shut down as their components failed. In 5 years the plant went from brand new to bandaids and concomitant with its demise came the closing of the facility.

With no plans or programs in place to manage the operation, it was doomed to fail from its inception. My only question is... what took it so long? That they last as long as they do is dubious praise for the capacity of our machines to withstand the abuse we often subject them to!

Touring The Facility

If someone were to ask you where your power plant was located, you're normal response might be: in the annex, at the basement level. In reality, the plant permeates the whole physical structure that makes up your facility. The systems whose core components are situate in the power "house" have tentacles that snake through the walls, floors and ceilings of your building(s). Auxiliary units are often housed in separate rooms, as are electrical distribution panels, air handling units and booster and recirculating pumps. The roof may support such items as cooling towers, exhaust fans, elevator penthouses, lightning arresters, stacks and vents. Part of the plant may even be represented under ground, in conduit, carrying everything from condensate generated by heated sidewalks and electricity for security lighting to hard wire hook ups between your local fire department and your fire annunciator panels. Whatever comprises your power plant, it will behoove you to trace it out, document its particulars and manage it, otherwise... well, you're about to see what I'm getting at. Let's take it a room or area at a time. We'll create an equipment inventory and proceed while jotting down some of the problems associated with it. Mr. Boldt, if you'd be so kind as to show us the way...

Figure 1-1.

THE BOILER ROOM

Whew! What happened to this place? It looks like a junk yard dog broke his chains and came tearing through here looking for meat. Let's see if any of this stuff is salvageable. Allen, what have we got?

System/Device	Problem
Steam Boilers	- oversized for the loads put on them
	- burner programmer/safety interlocks malfunction
	- furnace refractory burned out
	- breeching corroded
	- gas valve regulators chatter
	- feedwater preheater heat exchanger

	internals corroded
	- blowdown lines perforated along their length and drains too small to accept discharge
Steam Distribution System	- header safety valve leaks continually pressure reducing stations suffer sporadic pressure discharges
	- inadequate number/location of expansion joints
Condensate Return/ Make-up Feed Water System	- main disconnect for feedwater pumps located on fourth floor
	- Fisher valves to DA tank scaled up internally/gauge glass broken
Fuel Oil Transfer System	- pumps seized
	- system requires manual priming
	- underground storage tanks not registered or maintained
Chemical Treatment Transfer System	- pumps malfunction
	- lines plugged and leaking
Auxiliary Devices	- water softener regenerating clocks broken
	- water meters broken
	- sump pumps malfunctioning
	- coils in domestic water tanks blowing through
	- no trap maintenance program exists
	- valves/pumps leaking like a sieve (never repacked)
	- nothing tagged for area serviced
	- combustion air intakes blocked with debris

Figure 1-2.

THE CHILLER ROOM

No improvement here... same clutter, same neglect... just different equipment with different problems.

System/Device	Problem
Centrifugal Chillers	- one just "doesn't run!" - control panel covers missing - indicator lights burned out - pressure/temperature gauges broken/missing - no history of operation or repair documented
Auxiliaries	- chilled water/cooling tower pumps need repacked - lagging (insulation) torn/missing and lines not marked

- pneumatic control air compressor doesn't maintain pressure
- no back flow preventers in water lines
- motor starter disconnects internals exposed
- domestic water booster pumps unable to maintain pressure on upper floors

Figure 1-3.

THE MAIN ELECTRICAL ROOM

System/Device	Problem
Incoming Power Transformer	- PCB contaminated
Main Distribution Panel	- panels not marked for areas served - no testing/maintenance performed since installed
Fire Alarm System	- indicator lights missing or burned out - panels not marked/switch handles missing - smoke and heat sensors missing or disconnected throughout the system - no service agreement for maintenance/no testing of systems/no documentation
Emergency Power Generators	- room heaters not functional - units not tested or maintained - battery cable missing/cells dry - air intake louvers stuck open/seal felts missing

Diagram Of Efficiency

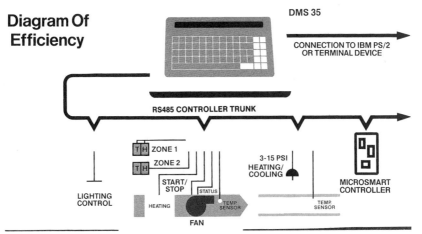

Figure 1-4.

THE MAIN AIR HANDLING ROOM

System/Device	Problem
Air Handlers	- vibration eliminators broken/missing - drive sheaves worn/improperly sized/out of alignment - heating/cooling coils leaking/plugged - drive motors running hot - magnehelics missing/broken - duct heaters disconnected - belts missing/worn/improperly sized - interiors corroded (out through jacketing) - history of frozen coils/freezstats bypassed - coil drip legs too short/condensate forced to run uphill - control wiring exposed/disconnected - humidification control valves not functioning - no trap program - insulation torn/missing - preheat valves stuck/frozen in open and/or closed position - units not maintained (obviously) nor marked for area served
Air Dryers	- no pad/mounted on bricks - puts out contaminated air/pressure fluctuates
Auxiliaries	- converter tube bundles leak - PRV's at steam station uninsulated and activate sporadically - louvre damper motors inoperable - sprinkler heads missing/wrong temperature rating

THE ROOF

Exhaust Fans	- covers missing/wires exposed/belts broken/motors seized/build-up on blades/bearings shot
Cooling Towers	- internals missing/corroded surfaces/drains plugged/fouling evident/no PM
Elevators	- penthouse dangerously cluttered/contacts burned/oil everywhere/units fraught with problems/operating certificates missing/no electrical schematics available/cables frayed/accessories missing
Lightning Arrestors	- disconnected

ANCILLARY ITEMS

Condensate Collection/ Transfer Tanks	- pumps inoperable (loss of condensate to drain)
Floor Drains	- positioned at high points allowing ponding
Fresh Air Intakes	- louvres broken/stuck shut and air flow impeded by clogged screens
Electric Motors	- slow running/operating single phased
Domestic Water Branch Lines	- once through/no recirculation in hot water lines
Valves/Piping	- improper zoning/numbers/location
Incinerator	- does not meet code
Structural integrity	- ceiling/floor/wall penetrations not filled in

Some Initial Considerations

So much for the cursory overview. I know we haven't seen it all, but we've certainly seen enough to establish our focus. It's not just the condition of the plant we've got to concern ourselves with. Power plant management entails more than equipment operation and repair. Other responsibilities include staffing and personnel issues, enforcement of departmental policies and procedures, creation and implementation of programs, equipment acquisition, installation and disposal, budget forecasting and engineering Rube Goldberg solutions to Murphy's Law difficulties... just to identify a few.

To that end, why don't we grab lunch and brainstorm the subject. We'll first compile an extensive list of things to consider, melt it down, then organize the renderings into an action plan. Here's my contribution...

Coverage of equipment (in-house/contractual) _____
Goals/objectives of department _____
Standard Operating Procedures Manual _____
Organizational structure _____
Water hardness/purity _____
Reliability of electric power/gas... _____
Number of sources of power _____
Number of sources of fuel _____
Redundancies in-system _____
Weak link scenarios _____
Depreciation schedules _____
Vibration analysis (predictive maintenance) _____
Fuel oil analysis (predictive maintenance) _____
Emergency power capabilities _____
Adequacy/purity of fuel supplies _____
Heat transfer efficiency _____
Nuisance tripping episodes _____
Corrective actions taken (past failures) _____
Asbestos management program _____
Lubrication history/program _____
Type/adequacy of insurances _____
Boiler/equipment lay-up _____
Performance ratings _____
Rules/Regs for operators _____
Licensure requirements _____

Repair/replace defective equipment (restoration) _____
Instrument calibration _____
Written rounds/routines _____
Noise control/spatial layout _____
Parts and consumables inventory _____
Manpower/Operating/Capital Budgets _____
Job descriptions _____
Identification/tagging of equipment _____
Work order system _____
Department hours of operation _____
Loading of equipment/systems _____
Reporting systems/documentation _____
Hazardous materials handling _____
Functional adjacencies _____
Steam trap program _____
Housekeeping/security _____
Safety valve testing/calibration _____
Instrument calibration _____
Equipment acquisitions/installation _____
Auxiliary equipment program _____
Alternating equipment operation _____
Overhaul/inspection schedules _____
Disposal of unwanted items _____
Vendor lists/purchasing procedures _____
Color coding of systems _____
Delineation of job duties _____
Communications with other departments _____
Technical library _____
Pressure pipe integrity _____
Condition of equipment components _____
Forms, logs and reports _____
Chemical storage/handling/transport _____
Insulation and equipment guarding _____
Energy conservation measures _____
Manufacturers safety data sheets _____
Emergency contingency plans _____
Inspection/testing frequencies _____
Pump packing _____
Manufacturers/blueprint indexes _____
Employee orientation/training _____

Air/moisture infiltration _____
Appropriate safety interlocks _____
Fire monitoring and protection _____
Structural stability (floor, walls, pads) _____
Sizing of equipment (new layout) _____
Proper air changes/filtration _____
Purity of control air/cooling water _____
Waste heat recovery _____
Adequacy of temperature/pressure ranges _____
System component winterization _____
Adequacy of expansion joint ducts/piping _____
Thermographic analysis surveys _____
Tooling up (hand/power) _____
Integrity of vibration eliminators _____
As-built drawings and close-out documents _____
Competency of staff/diversity of trades _____
Proper grounding of equipment _____
Regulatory compliance (federal, state, local) _____
Code compliance (NFPA, NEC…) _____
Registration of underground tanks _____
RFP's for service contracts _____
Equipment repair histories _____
Preventive maintenance schedules _____
Scheduled shut-downs _____
Start-up/operating/shut-down procedures _____
Water treatment program (boiler/cooling tower) _____
… care to add a few of your own?

_____ _____
_____ _____
_____ _____

Tentative Implementation Plan

Why tentative? Because at this juncture, the only thing you can be certain of is that your plan will be altered innumerable times as you implement it. The following sample plan might help you establish a format and the first chapter may jostle some pertinent considerations from your head but nothing can substitute for the hands-on, eyes-open, ears-cocked, sniffing you'll need to do on frequent walk-through investigations of your plant.

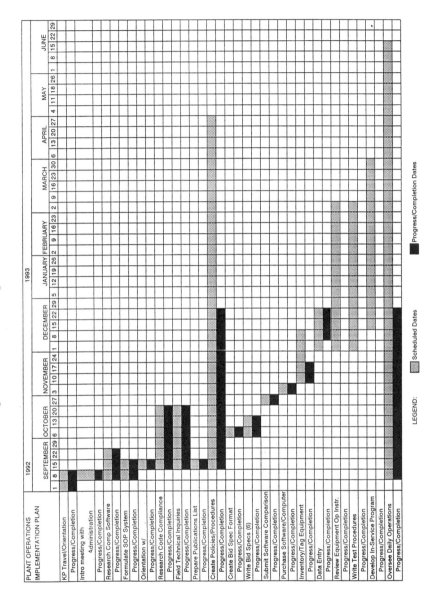

Figure 1-5. Master Implementation Plan

Chapter 2

Standard Operating Procedures

Just as canon laws are enacted by religious groups to insure the propagation of their beliefs and playbooks are constructed by coaches of organized sports to guide their teams onward to victory, so too must facilities, and their management conform to established sets of rules if they are to succeed and endure. In my experience, the corporate bible (otherwise known as the SOP Manual), forms the basis for all future organizational failures or successes. Depending on how detailed an instrument it is, how well organized, how frequently utilized and the currency of the information it contains, it can literally make or break your operations.

Setting up a System

I cased the place in hopes of exhuming an old SOP Manual we could utilize as base documents in instituting a new system but could come up with only a few ill-conceived, rudimentary directives such as "... uniforms must not be too large or too small"... DA! and "... do your wirk in the general direction of the movement, such as moping the flor." Gee, I wonder why they went belly up so quickly?

So that our organization doesn't suffer the same fate, I put together a 20 volume "system" of policies and procedures. Not just a collection of binders relegated to a shelf and used for filing company memos, it is a living instrument designed for daily use and reference in implementing our programs, documenting our compliance with applicable standards, and exacting strategies for future successful operations.

Figure 2-1
SOP Manual System Outline

PLANT MANAGEMENT	
POLICY TITLE:Forward/Standard Operating Procedures Manuals	POLICY NUMBER: 3-06.1.1.0
DATE OF ORIGIN: 10/9/92 REVIEWED & REVISED	FOR DIVISION USE PAGE 1 OF 5 FOR DEPARTMENT USE X

I. PURPOSE:

To define the roles of and standardize policy and procedure guidelines for coping with the diverse programs established within departments under the direction of the Director of Facilities Management and provide coordination of and direction for their operations and staff.

II. ORIGINATOR:

Director of Facilities Management

III. SCOPE:

Plant Maintenance and Facilities Maintenance

APPROVED:

    (Signature)   (Title)   (Date)

Policy #:3-06.1.1.0

IV. TEXT

A. Format:
All departmental policies/procedures will follow the format established by this policy and the following referencing method will be utilized to provide a logical organization of each P & P subsequently published, included in and adopted for use in the manuals.

Every policy and/or procedure will be assigned a unique number that cannot be duplicated based on this system of coding:

(Example 1. 4. 6. 2)

Whereas:
- The 1st number in the string (1) designates the department ID.
- The 2nd number in the string (.4) designates the area of interest.
- The 3rd number in the string (.6) designates the topic policy.
- The 4th number in the string (.2) designates a procedure referencing the topic policy.

1) Department Identification
Following is an inventory of department ID's based on these cost center numbers, presently in force.

Cost Center #	Department
3-01	FACILITIES & SUPPORT ADMIN
3-02	FOOD SERVICES
3-03	HOUSEKEEPING SERVICES
3-04	MATERIAL MANAGEMENT
3-05	PURCHASING
3-06	PLANT MANAGEMENT
3-07	PLANT MAINTENANCE
3-08	FACILITIES MAINTENANCE
3-09	MAIL DISTRIBUTION
3-11	SECURITY SERVICES
3-15	COMMUNICATION SERVICES

Policy #:3-06.1.1.0

2) Area of Interest

The Standard Operating Procedures manuals is a multi-faceted document comprised of 20 volumes covering as many management areas, which encompass the sum total of the accountability of all departments that fall within the parameters of the Plant Management Department.

Currently the areas addressed include:

Volume	Management
01	Department Operations
02	Financial/Contracts
03	Personnel Related
04	Training/In-Service
05	Work Order System
06	Equipment Management
07	Preventive Maintenance
08	Safety Management
09	Fire Protection
10	Open
11	Emergency Preparedness
12	Service Agreements
13	Regulatory Compliance
14	Energy Management
15	Committee Membership
16	Construction/Renovation
17	Quality Assurance
18	Open
19	Open
20	Open

3) Policy Topics
Subject matter addressed for which policy is being established

4) Procedure
Actions to be taken in support of an established policy.

Policy #:3-06.1.1.0

B. Policy Development/Coordination

   Policies/procedures shall be written by all departments in establishing their respective roles within the organization. The cooperation of other department managers affecting or affected by such policy/procedure should be solicited prior to its construction; as should representatives of committees, and administration as appropriate.

C. Usage

   P & P's shall be developed and maintained by and in each department to:

   1. Establish a basis for evaluating the correctness and effectiveness of operating parameters, methods and techniques.
   2. Standardize practices and processes throughout the facility.
   3. Provide a ready reference for employee and/or outside agencies requiring such information.
   4. Serve as an aid in the orientation of personnel foreign to the operations covered.

D. Availability/Location:

   A current set of P & P manuals, complete with index and updated revisions, must be prominently displayed within each department. The department manager is responsible for making said manuals readily available for reference by all department personnel.

E. Revisions/Approvals

   Prior to inclusion in the P & P manual set, all newly written and/or revised policies/procedures must be initialled by the Director of Facilities Management before being forwarded on to Administration for final approval.

Section Modularity

The number of sections comprising the "system" needn't be 20; it can be any number you desire, limited only by your imagination and/or operating requirements. I chose 20 because that value fulfills our present needs while providing the flexibility for future inclusions should they become necessary.

The system's modular construction enables the utilizer to focus on particular aspects of the operation individually and allows the easy insertion and extraction of relevant information into and out of each section.

There needn't be rhyme nor reason to the order of the sections: they can be installed alphabetically, in finance, equipment, activity of personnel intensive blocks or by any other design. The order mine are in happen to be a consequence of when I addressed each area. Personally, I would recommend against the alphabetical approach due to the problems entailed in renumbering your policies should you include a new section into the core of your system later on.

Typical Policy Topics

Whole volumes can be compiled listing the infinite possible topics for which policies can be written. My purpose here isn't to instruct you as to which one you should address; that will be determined by the nature of your operations, jurisdictional authorities to which you are subject, requirements of the organization for which you work and the level of commitment you bring to your job. Rather, I offer you the list of those I've mustered for this place, for your use as a starting point in constructing your own. Feel free to massage it in suiting it to your own purposes:

AREA OF INTEREST	POLICY TOPIC
Department Operations	—Mission Statement
	—Organizational Charts
	—Goals/Objectives
	—Hours of Operation
	—Performance Standards
	—Rounds/Routines
	—Winter Groundskeeping

AREA OF INTEREST	POLICY TOPIC
	—Summer Groundskeeping
	—Job/Shop Clean-up
	—Schedules and Forms
	—Blueprint Index
	—Management Objectives
	—Department Duties
	—Overnight Accommodations
	—Flying the Flag
	—Construction/Renovation
	—Minor Projects
	—Dispatch Log
	—Handicap Access
	—UPS Systems
	—Tool Care/Use
	—Staff Meetings
	—Master Equipment List
Financial/Contracts	—Man Hour Budget
	—Capital Budget
	—Operating Budget
	—Purchase Order system
	—Equipment Acquisition
	—Request for Proposal
	—Request for Quotation
	—Water Softener Salt
	—Rock Salt
	—Inventory Control
	—Purchased Services
	—Blanket Purchase Orders
	—Cost Center Charge Backs
	—Vendor Relations
	—Resource Control
	—Prepurchase Evaluation
Personnel Related	—Interviewing/Hiring
	—Employee Handbook
	—Job Descriptions
	—Dress Code/Conduct
	—Performance Evaluation

AREA OF INTEREST	POLICY TOPIC
	—Timekeeping/Records
	—Job Posting
	—Personnel Roster
	—Licenses/Renewals
	—Memberships
	—Work Schedules
	—Vacation/Holiday/Sick Time
	—On-Call/Call-in Procedure
	—Staffing Parameters/Levels
	—Overtime Scheduling
	—Grievance Procedure
	—Absenteeism/Tardiness
	—Work Assignment
	—Hand Tool Distribution
In-Service/Training	—Employee Orientation
	—Mandatory In-service
	—Seminar Attendance
	—Tuition Reimbursement
	—Department Training
	—Annual Retraining
	—Training Records
Work Order System	—Maintenance Cart Rounds
	—Requisitioning of Work
	—Program Components
	—Project Work Request
	—Work Order Follow-up
	—Corrective Maintenance Guide
Equipment Management	—Equipment Operating Manuals
	—Freeze Protection
	—Generator Operating/Testing
	—Start Up/Shut Down
	—Equipment Lay Up
	—Scheduled Shut Downs
	—Overhauling Equipment
	—Filter System
	—Valve/Switch Lists

AREA OF INTEREST	POLICY TOPIC
	—Plant Operating Logs
	—Repair Procedures
	—Electrical Distribution
	—Moving of Equipment
	—Alternating Operation
	—Power Factor Correction
	—Inspection/Testing
	—Certificates of Operation
	—Elevator Traffic Plan
	—Water Treatment Reports
	—Valve Tagging
	—Incinerator Cleaning
	—Pneumatic Tube System
Preventive Maintenance	—Program Computerization
	—Mechanical Rounds/Routines
	—Groundskeeping Equipment
	—Vehicle Care
	—Humidity Control Guidelines
	—Equipment PM Histories
	—Consumables Inventories
Safety Management	—Electrical Safety Program
	—Department Safety Manual
	—Lightning Protection
	—Portable Space Heaters
	—Microwave Oven Testing
	—General Safety Rules
	—Safety Inspections
	—Extension Cords/Adapters
	—Risk Management Program
	—Securing Mechanical Rooms
	—Key Control
	—Electrical Receptacle Tests
	—Hot Water Temperatures
	—Faucet Aerator Maintenance
	—Welding Procedures
	—Storage Room Clutter

AREA OF INTEREST	POLICY TOPIC
	—Roof Access
	—Protective Gear
	—Eye wash Stations
	—Lock Out/Tag Out Procedures
	—Hazardous Materials Program
	—Accident/Incident Reporting
Fire Protection	—Smoking Policy
	—Fire Drills
	—Fire Extinguisher Inspection
	—Stand Pipe Testing
	—Fire Department Tie-in
	—Hazardous Materials Program
	—Trash Disposal
	—Use/Storage of Flammables
	—Fire Brigade Duties
	—Fire Pump Testing
	—Duct Fire Dampers
	—Building Compartmentalization
	—Fire Alarms/Detectors
	—Management of HVAC
	—Automatic Sprinkler System
Emergency Preparedness	—Interruptions of Utilities
	—Equipment/System Contingencies
	—Loss of Fuel Supply
	—Bomb Threat
	—Telephone Outage
	—Loss of Electrical Service
	—Emergency Plans/Procedures
	—Snow Management Plan
	—Emergency Work Schedule
	—Loss of Water Supply
	—Call in List
	—Emergency Vendor List
	—Severe Weather Plan
	—Disaster Cabinet
	—Emergency Lighting
	—Essential Services

AREA OF INTEREST	POLICY TOPIC
	—Evacuation Procedures
	—Disaster Plan
Service Agreements	—Boiler Flame Safeguard
	—Chiller/Cooling Towers
	—Elevator Service
	—Water Treatment Program
	—Minor Refrigeration Equipment
	—Fire Warning System
Regulatory Compliance	—Asbestos Management
	—PCB Contaminated Devices
	—Underground Storage Tanks
	—Stack Emissions
	—Refrigerant Reclamation
	—Use of Biocides
Energy Management	—Outdoor Photocells
	—Energy Conservation Measures
	—Computerized Management
	—Evening Set Points
	—Peak Shaving
	—Power Factor Correction
	—Repair of Insulation
	—Free Cooling
Committee Membership	—Attendance Log
	—Meeting Minutes
	—Committee Charters
	—Membership Structure
	—Meeting Structure
	—Issue/Action Reports
Quality Assurance	—Boiler Plant Inspection
	—Mechanical Room Inspection
	—Public Area Inspection

CHAPTER 3

CONTRACTED SERVICES

Good morning (Mr. Phelps). We are faced with a controllable dilemma. As this project is considered neither new construction nor a renovation but rather an attempt to make a dilapidated facility whole again, the physical plant equipment and systems are being turned over to us as they're restored. We are staffed with a skeleton crew of operating engineers and no PM program. Our mission (whether we like it or not) is to maintain the power plant until a full complement of personnel have been hired and a computerized maintenance management program has been installed. As always, if either of us is caught or killed, we obviously will have done something dreadfully wrong. This message should help to keep you from self destructing.

Pertinent Focal Points

Sooo... what are we looking to accomplish? After all, the only reason this stuff is our responsibility is because it's been totally restored and is 100 percent functional. Hell, there's even a parts and workmanship warranty attached to most of it. But in some cases the warranty is only for 90 days and coverage began the minute it was turned over to you (little solace when it's a chiller in December or boiler in July). Besides, parts may refer only to broken parts, not malfunctions, and you've got to keep this place running—READ THE CONTRACT!

Once you've perused and appropriately filed it in the round receptacle, you'll better appreciate your elders' contention that you "...get nothing for nothing." But what you will get, if you don't provide for adequate care and repair of your equipment and systems, is headaches and telephone calls in the middle of the night. With that in mind, let's go over the Master Equipment List we compiled earlier and develop a strategy for maintaining it in proper working order: starting in the boiler room...

Granted, we have a cadre of operating engineers working here, but the definitive word here is "operating." Their machinery repair experience hasn't been documented nor have their mechanical skill sets been determined. Rather than challenge them with an immediate reversal of the way they are accustomed to doing their job, I suggest that we supplement their efforts with outside help until they've been subjected to proper and adequate training and/or the department is fully staffed and a maintenance/repair capability has been established. Initially, consideration should be given to contracting for the calibration of the flame-safeguard control systems, make-up water analysis/treatment and preparation of the boilers for their annual operating certificate inspections. The responsibility for maintaining the auxiliaries should be turned over to the in-house people; one, to begin the skill assessment process and two, to serve as a starting point for the implementation of the preventive maintenance program that comes later.

As we progress out into the plant, other facets of the operation we may deem necessary might include fan/system balancing, equipment vibration analysis, treatment of cooling waters, thermographic studies of the roof/electrical systems, etc.; but before the appropriate vendors can arrive to begin work, we must provide a proper vehicle to get them here.

Proposals Versus Quotations

What the hey... you would have had to deal with them sooner or later anyway, so you might as well put pen in hand and start writing. Beyond your organizations purchased services policies (3-bid minimum?) and/or the budgetary constraints you've had imposed on you (I've got a blank check!), there are basically two ways to go about acquiring service agreements on your systems/equipment... PROPOSALS OR QUOTATIONS.

Before I go any further, I want to clear up a long unresolved misunderstanding I've had with several people over the years regarding what the difference is—and there is definitely a difference. To my mind, a request for proposal is a formal procedure utilized to determine what products, services options, etc. are available from and deliverable by expert companies/consultants at what frequencies/amounts, etc., over which periods of time. You needn't know all the intricacies involved beforehand, only that you have an unfulfilled need. It's important that you send each vendor identical information to consider but each of the contract proposals you receive back will be comparatively different based on many factors, individual to the responding bidders. In other words, they will have *proposed* what they are able and/or willing to provide you, for a given duration and dollar amount. At that juncture you might accept one of the bids, have them all rebid based on new information or renegotiate for better terms before signing a contract.

Figure 3-1. RFP Skeleton

Vendor Qualifications

Bidders must provide sufficient information regarding each of the below listed criterion to enable the owner to thoroughly evaluate their firms ability to professionally service the account.

- years in this business
- minimum of five local references from comparable type and size facilities
- adequacy, competency and availability of staff
- licenses, certifications, and insurances carried and current
- technical depth of your organization
- ability to acquire parts
- narrative of how you plan to service the account
- principals of record

Vendor Responsibilities

Bidder must agree to each of the below listed criterion; and make appropriate comments regarding them in the body of their proposal documents:

- conduct site investigations as needed
- call attention to owners omissions, regulatory compliance issues and documents/drawings required by the vendor, necessary to the bidding process

(Continued)

- provide a list of contacts and a procedure to follow for emergency service
- apply for and acquire all required permits and licenses
- submit a plan for administering the agreement
- install a documented training program for the owners employees
- properly dispose of all wastes generated
- follow the owners vendor control procedures as established
- address these areas as appropriate:
  - provisions for loaners
  - troubleshooting of problems
  - compilation of inventories
  - vendor employee disputes

Definitions
Bidder shall provide a section in the proposal which defines their interpretation of the following terms as applicable.

- level of service
- call back
- call in
- emergency
- normal business hours
- premium time
- portal charge
- covered parts
- cost of services
- hourly rates
- extras

Terms
All bids are to be based on a one-year agreement. Longer contract lengths may be cited separately as options. Items the bidder must address in this section are:

- length of contract
- cost of services
- renewal procedures

- contract termination
- additions/deletions
- price increases

Stipulations
In addition to the foregoing, the owner reserves the right to:

- determine the adequacy of substitutions
- require specific response times
- approve sub-contractors before hiring
- forego the use of reconditioned parts
- expect calibration of vendor test equipment to National Bureau of Standards requirement
- receive all correspondence from the vendor in writing, signed, and dated
- have stocking levels maintained by the vendor that will assure timely consummation of the agreement

Work Scheduling
All requests for site visits for information gathering prior to bid must be made through the office of the Director of Facilities Management. Upon receipt of a signed contract, the vendor shall contact the Chief Engineer to coordinate scheduled shut downs, arrange for vendor access to the premises and establish mutually agreeable hours of coverage and service intervals.

Documentation
Vendor shall supply owner with complete documentation relating to all aspects of the service area including, but not limited to:

- Manufacturers Safety Data Sheets
- signed service reports
- warranties/guarantees
- repair histories
- planned (pm) repair schedule
- calibration documentation
- test results
- laboratory analysis results
- detailed procedures

(Continued)

> - operating manuals
> - drawings/schematics
> - equipment specifications
>
> **Summary**
> After reviewing the foregoing specifications, bidders shall provide the owner with an overview of their proposed program, citing their ability to do the work and indicating the earliest date they can start.
>
> Warning: Failure to provide any of the information called for in the specs will place your bid in jeopardy of disqualification
>
> **Addendum**
>
> _____

On the other hand, as an expert who knows exactly what products/services are required, you may want to write up sets of instructions and/or specifications which detail their delivery in exacting dimensions, quantities and time frames, leaving nothing to the imagination. In a request for quotation, the specifications you provide are incorporated into a bid package that stipulates the scope, terms, conditions and length of the contract to be let. The RFQ is a restrictive instrument, in that it limits the number of vendors who can bid on the contract based on its inflexible nature and doesn't account for critical items left out during its construction. Again, as with an RFP, it's important (even more so) that you provide the same information to all the bidders, but the only difference you'll see in their responses will be in the "proposed" cost of the contract. In other words, they will have *quoted* you a price to provide you with exactly what you requested.

Figure 3-2. RFQ Skeleton

SPECIFICATIONS

(Overview of equipment to be covered, products or services to be provided and/or areas addressed)

DESCRIPTION OF WORK

(maintenance, inspection, repair, adjustment, safety testing, replacement, calibration... etc.)

SCOPE

(minimum requirements, code compliance, acceptable engineering methods/techniques... etc.)

BIDDER QUALIFICATIONS

(References, years in business, doing comparable size/type work, adequacy of staffing, technical skill level of employees, permits/licenses/certification, in-service/limits required, technical depth, principles of record, ability to acquire parts)

BIDDER RESPONSIBILITIES

(site investigation, call attention to owner omissions, acquisition of permits, required documents/drawings, training of owners employees, disposal of waste; clean up, compile inventories, provide loaners, troubleshooting, follow hospital vendor control procedures, character of work)

DEFINITIONS

(level of service, call-in, call-back, emergency, normal business hours)

TERMS

(length of contract, cost of services, premium time, portal charges, renewal procedures, normal business hours, additions/deletions

(Continued)

to contract, expected performance, hourly rates, holiday after hours coverage, handling of price increases/extras, covered parts, itemized invoicing)

STIPULATIONS

(Required response times, adequacy of substitutions, right to approve, sub-contractors, use of reconditioned parts, on-call availability, calibration of vendor test equipment to NAT'L BUREAU OF STANDARDS requirements, all correspondence in writing/dated/sized, warranties/guarantees, stocking levels, in-services for affected personnel, contract needs all specs... etc.).

WORK SCHEDULING

(Coordination of scheduled shutdowns, normal hours of coverage, vendor access to premises, vendor contact, established service intervals)

DOCUMENTATION

(MSDS sheets, service reports/histories, planned PM/repair schedules, calibration documentation, laboratory analysis and test results, detailed procedures used, operating manuals, drawings)

ADDENDUM

(Equipment inventories, list of loaner availability required, spare parts and materials, special equipment/procedures, consumables inventory... etc.)

I'm sure you can envision when one process might be used to better advantage than the other (even for the same function) depending on the circumstances involved. But in an effort to help you mind your (RF)P's and (RF)Q's, allow me to offer the following bid request for servicing of our elevators.

So... are we asking for a proposal or a quotation? In either event, you got yourself a pre-written bid spec for servicing of your elevators.

Figure 3-3. Bid Specification

ELEVATOR MAINTENANCE

(FULL SERVICE MAINTENANCE AND REPAIR AGREEMENT)

SPECIAL CONDITIONS

SC-01 SCOPE

To furnish supplies, parts, materials, equipment, labor and shop facilities as called for in the contract documents, specifications and proposal sheets.

SC-02 RUBBISH

The contractor shall at all times keep the premises free from accumulation of waste materials and rubbish by his employees of work and at the completion of work, he shall remove all his tools, surplus material, etc., and shall leave the premises and his work in a clean and orderly manner.

SC-03 INSURANCE

Upon award of the Contract, the Contractor shall procure and supply to the Purchasing Department, Certificates of Insurance describing his/her policies of Insurance. Such Insurance shall include, but not limited, Workmen's Compensation and Employer's Liability Coverage with minimum limits of $1,000,000.00 per person, per accident. Comprehensive General Liability, including Broad Form, Contractual, Product Liability, et al, coverage with minimum limit of $5,000,000.00 for Bodily Injury, $2,000,000.00 for Property Damage and Personal Injury. Auto and Property Damage in aggregate. Excess Liability, Umbrella Form Coverage with minimum limits of $5,000,000.00.

(Continued)

Should any of the above described policies be cancelled before the expiration date thereof, the issuing company must provide thirty (30) days written notification prior to such cancellation. This should be noted on Certificate of Insurance. Certificate of Insurance shall be submitted before the performance of service.

SC-04 SUBSTITUTIONS

In the event of material substitutions, where permitted, properly identified samples must be available to the Department of Purchase upon request. In addition, it is fully understood that the Purchasing Agent reserves the right to select the product(s)

SC-05 CONTRACT PERIOD

This is a full service maintenance and repair type contract for a period of three (3) years effective after approval by the Purchasing Agent and after proper execution of the Contract Documents and Performance Bond. The Contractor is to start within fifteen (15) days after the award. The Performance Bond shall be renewed annually.

SC-06 TERMINATION

All Contracts are subject to cancellation upon written notice, allowing thirty (30) days notification for termination of such contract by the Purchasing Agent.

SC-07 LABOR RATES

The prevailing rate is to be paid to the Mechanics/Laborers for the work performed as prescribed by the City and County Building Trades, for the duration of this contract.

SC-08 PERFORMANCE BOND

There shall be a Performance Bond required for the purpose of this proposal in the full amount of the Contract.

SC-09 PERMITS

Contractor shall assume all responsibility and expense for obtaining any and all permits, fee or non-fee, and shall pay all taxes required in complying with City, Local, County and State Law, Codes or Ordinances. This shall be in effect on this Contract (Purchase Order) and Labor Requirements between the Contractor and the county of _____.

SC-10 GENERAL CONDITIONS

1. The Contractor shall be required to have and substantiate five (5) years experience in repair and maintenance as specified and must have a fully equipped service center with complete shop and parts inventory.

2. The Contractor shall be responsible and make every effort possible to maintain said elevators under this contract; in order, that no elevator shall be in violation of the present code or changes to the code for the duration of this contract. The present code as used by the City of Chicago, Department of Inspectional Services, Elevator Bureau - 79.1 includes 1971 American National Standards Institute. (ANSI) A17.7, 1971 with supplements AN17.1A, A171C1974, A17d1975, A171F1975. The above or latest edition in effect during the time of this contract shall be the code used in all cases, including supplements.

SC-11 DEFINITIONS

A. Regular Working Hours: Monday through Friday - 7:00 a.m. to 4:30 p.m. Excluding Nationally Observed Holidays.

B. Over Time Hours: Anytime after or before normal working hours.

C. Full Maintenance Call Back: Call back after regular working hours at no extra charge.

(Continued)

D. Full Maintenance Time Differential Call back: Call back after regular working hours shall be charged at the rate of the premium portion of the overtime charge.

1. Where call back service provisions are stated, it is to be understood that this provision applies only to calls for services on the portion or portions of this contract that is awarded as an 8 hour service see (A) & (C) above, but shall include all materials and parts.

2. Further, the portion awarded as full 24 hour service the price quoted is to include all parts and materials as well as call back service, see (B) & (D) above.

E. The Contractor's personnel, upon each service call shall check in with the Chief Engineer at the site prior to the performance of any work. Subsequently once the work is completed, the contractor's employee shall complete a job sheet detailing work done, material used, and the time spent to accomplish the work. This job sheet shall be signed and approved by Chief Engineer and be submitted concomitantly with the invoice forms 29 or 29A.

SC-12 PROVISIONS

1. The End of Contract Privilege: Ninety (90) days before the end of this contract the owner shall reserve the right to have any or all elevators listed in this contract inspected by a third party. The third party shall inspect and document by Punch list any and all items covered under this "Full Maintenance Program", as defined herein, that shall be corrected or replaced by the Contractor at no cost.

2. Repair service will typically be available within four (4) hours after request, subject to acts of God and circumstances beyond Contractor's control. The contractor shall make every effort to service he equipment as quickly as possible. When notified of an emergency repair (patient

or passenger in jeopardy) or replacement; service calls shall be supplied in a time not to exceed one (1) hour after notification. Failure to comply may be taken as just cause for cancellation of this contract by

3. All defective or questionable parts of materials unless specifically excluded shall be replaced at no extra cost. Replaced parts which are repairable or are normally sold on an exchange basis shall be the property of the Contractor. Replacement parts of materials shall be exactly the same manufacturer, model and part number as those removed; unless, certification as to equal or better operating and functional parameters are furnished to the hospital giving name, model # and serial # of equipment parts. This will include updating and modifications as necessary.

4. The Contractor is authorized to purchase "spare" parts for the equipment in advance in the name of the owner as required for the maintenance ad repair of the equipment. In each instance that this authority is utilized, the Contractor shall indemnify and hold the owner harmless from any and all liability that may arise from the use of such authorization. The adoption of this provision is intended to expedite and improve the services of the Contractor to perform under this contract with the owner and will be used only for this purpose.

SC-13 INSPECTION/WORK LOGS

The method of Inspection, Logging of Work accomplished and any other necessary record keeping systems for the administration of this Contract will be mutually agreed upon by owner and the contractor.

SC-14 VANDALISM/ABUSE

If the Contractors personnel, upon examination, ascertains that the damage and malfunction to the elevator or escalator is the

(Continued)

direct result of abuse of vandalism; then this information shall be reported to the Supervisor/Engineer on duty prior to the Performance of the service repair. The Engineer on duty will determine the schedule for said repair. The owner will not be charged for the service call other than the premium portion of Full Maintenance Time Differential if such elevator/escalator is covered under the Full Maintenance Time Differential price of the Contractor, (See Paragraph C & D under Definitions) plus cost breakdown sheets.

SC-15 EXCLUSIONS

The service provided by the Contractor dos not include:

1. Repair of damage resulting from catastrophes such as fire, flood or any act of God.

2. Repair of damage resulting from accident, neglect, misuse or operator abuse.

Proof of the above Paragraph shall be furnished by the Contractor to owner.

SC-16 GENERAL

A. The equipment covered shall be owned by the customer -

B. The Contractor shall have full and free access to the listed equipment, in order, to provide the contracted service with the time scheduled, and the approval of the Chief engineer for said work.

C. The Contractor is not responsible for failure to render service due to happenings beyond his control.

D. The Contractor shall keep an "Inspection Maintenance, and Repair Summary" on all equipment serviced under the Contract, employing a record format already acceptable and compatible to owner. The service records shall be current at all times, and shall be maintained at the Using Department for twenty-four (24) hour availability and reference.

E. The availability of Contractor's Service Facilities for Major Repair shall be provided when mutually agreed upon that such service is necessary/required.

SC-17 CLEANING

All scrap, rubble, debris, and defective parts or fixtures shall be removed from the building, and hauled off the site at the end of each working day. The owner's refuse containers "will not" be used for debris disposal.

SC-18 SITE INSPECTION

The contractor "shall" visit the site of the proposed work and thoroughly familiarize himself with the locations, the operating conditions and the conditions he will encounter affecting the proposed work. No additional allowance will be granted because of lack of knowledge of such conditions. To set an appointment contact:

Record date and time of inspection, areas inspected, thoroughness of inspection. Submit copy of inspection report to Purchasing for contract filing.

SC-19 AWARD OF CONTRACT

The Contract award shall be made only after an inspection of the bidder's Service Center Shop facility, parts, inventory and shall be contingent upon a finding of its adequacy for the performance of all maintenance and major repair support functions. A finding of adequacy by two of the three members of a team consisting of a Department of Purchase Specifications Engineers, Building, & Grounds Department, as representative of the owner shall constitute a finding of adequacy within the meaning of this provision. The findings of the aforementioned inspection team shall be re-

(Continued)

duced to writing and shall in addition to detailed finding of capacity equipment, personnel and proximity, include the general finding of adequacy or lack thereof.

SC-20 BID DEPOSIT

In accordance with the General Requirement of Bid deposit, a Bid Deposit in the Amount of _____ will be required and must accompany the Bid Proposal as Check, Bank Cashier's Check, Certified Check, or Band Treasurer's Check. Said Bid Deposit shall be made payable to the order of _____ Bid Bond will be accepted in lieu of said Deposit providing the Surety Company has an A+ rating and minimum classification of class VII.

SC-21 EQUAL EMPLOYMENT OPPORTUNITY REPORT

The undersigned acknowledges the requirement for filing a completed Contract Compliance Report Form #1 regarding Equal Employment Opportunity and certified that:

( ) Form #1 enclosed with Proposal

( ) Form #1 previously submitted and on file for

SC-22 CERTIFICATE OF QUALIFICATION

Contractor shall submit three (3) copies of Certificate of Qualification with his Proposal. Failure to fill out the said form properly may subject bidder to disqualification.

SC-23 REMOVAL FROM BIDDING LIST

The Purchasing Agent reserves the right to remove any Contract from its list of Acceptable Bidders who fail to comply with the requirements and specifications of these Contract Documents.

SC - 24 INQUIRIES

For further inquiries please contact:

SPECIFICATIONS

VERTICAL TRANSPORTATION MAINTENANCE SPECIFICATION

<u>1. DUTIES OF CONTRACTOR</u>

Contractor shall furnish all supplies, materials, labor, labor supervision, tools, equipment and lubricants necessary to provide full-preventive maintenance, adjustment, replacement and repair service for the complete vertical transportation system described below:

LOCATION:

<u>EQUIPMENT DESCRIPTION:</u>
(Main Building)

Car No.	Manufacturer	Capacity
S-1	Dover	4500 lbs
S-2	Dover	4500 lbs
S-3	Dover	4500 lbs
RO-1	Dover	3500 lbs
RO-2	Dover	3500 lbs
DW	D.A. Matot	500 lbs

(Building)

Lift   (single)

SPECIFICATIONS

<u>2. TERM</u>
The term of this contract, as indicated in the Special Conditions, shall be subject to the following:

(Continued)

If during the term of this agreement, Contractor violates any of the provisions of this contract or fails to properly provide the services required by this contract, Owner shall advise Contractor of specific deficiencies and shall allow a reasonable period (30 days unless otherwise agreed) to correct these deficiencies to Owner's satisfaction.

In the event Contractor fails to correct the deficiencies in the allotted time, the Owner shall have the right to terminate this agreement on 30 days' written notice to the Contractor.

If the Owner chooses to modernize operational and/or motion controls on vertical transportation equipment, contract may be cancelled with 30 days' written notice.

3. INSURANCE

Contractor shall secure and keep in force during the life of this agreement at his sole cost and expense, insurance policies (or the equivalent thereof) in companies acceptable to Owner as required by the Special Conditions.

Nothing in this agreement shall be construed to mean that Contractor assumes any liability on account of accidents to persons or property except those directly due to negligent acts or omissions of Contractor, its employees, subcontractors, servants or agents. Contractor shall not be held responsible or liable for any loss of damage due to any cause beyond its control, including, but not limited to, acts of government, strikes, lockouts, fire, explosion, theft, floods, riot, civic commotion, war, malicious mischief or act of God, nor in any event for consequential damages. Dates for the performance or completion of any ongoing maintenance or corrective action required by Section 2, Paragraph A, shall be extended by such length of time as may be reasonably necessary to compensate for the delay.

4. HOURS AND MANNER OR WORK
All normal work except as otherwise noted under this agreement including unlimited emergency call-back service will be per-

formed during regular hours of regular working days, as defined in the Special Conditions. Note: requirement for 7 day, 24 hour callback on selected elevators as indicated in description equipment.

SPECIFICATIONS

Removal of elevators from service shall be coordinated with and approved by the Owner or this Representative; Owner agrees to permit Contractor to remove elevators from service for a reasonable time in order to perform maintenance thereon:

Contractor shall provide a minimum of one resident mechanic and one resident helper on-site between the hours of 8:00 a.m. and 4:30 p.m., Monday through Friday, with the exception of trade-recognized holidays, for the exclusive purpose of performing preventive maintenance and providing emergency call-back service.

All repair work shall be performed by crews other than the resident mechanic and helper.

5. OWNER'S RIGHT TO INSPECT AND REQUIRE WORK

Owner reserves the right to make such inspections and tests whenever necessary to ascertain that the requirements of this agreement are being fulfilled. Deficiencies noted shall be promptly corrected at Contractor's expense.

If Contractor fails to perform the work required by the terms of this agreement in a diligent and satisfactory manner, Owner may, after 10 days' written notice to Contractor, perform or cause to be performed all or any part of the work required hereunder, Contractor agrees that it will reimburse Owner for any expense incurred therefor, and Owner at his election may deduct the amount from any sum owing Contractor. The waiver by Owner of a breach of any provision of this agreement by Contractor shall not operate or be construed as a waiver of any subsequent breach by Contractor. A qualified Elevator Consultant acceptable to both parties may be retained by Owner to mediate any disputes.

(Continued)

6. CONTRACTOR TO COMPANY WITH LAWS

In the performance of this contract, the Contractor agrees to abide by all existing laws, codes, rules and regulations set forth by all appropriate authorities having jurisdiction in the location where the work is to be performed.

Contractor shall make periodic tests and maintenance inspections of all equipment as required by current applicable safety codes for elevators, dumbwaiters, escalators and moving walks, including but not limited to, annual no-load, slow-speed test or car and counterweight safeties, governors and buffers; annual pressure test on hydraulic elevators; and 5-year, full-load, full-speed, test of safeties, governors and buffers; al as required by ANSI A17.1 Code. Written reports of said tests shall be submitted to the Owner and, in the Cage of running safety tests, prior notification shall be given so that a Representative of Owner may witness said test.

Under this agreement, the Contractor shall not be required to install new attachments or perform tests other than those specified herein as may be recommended or directed by inspecting entities; insurance companies; and federal, state, or municipal governmental authorities subsequent to the date of this contract, unless compensated for such installation or services.

SPECIFICATIONS

7. EMPLOYEES OF CONTRACTOR TO BE SATISFACTORY

Contractor agrees that all work shall be performed by and under the supervision of skilled, experienced, elevator service and repair persons directly employed and supervised by Contractor. Any and all employees performing work under this contract shall be satisfactory to Owner.

8. EXTENT OF THE WORK

Contractor shall be responsible for regular, systematic execution of the work items included in this contract as follows:

<u>Complete Maintenance</u>: Contractor agrees to regularly and systematically examine, clean, lubricate, adjust the vertical transportation equipment and provide emergency call-back service per Section 5 of this agreement, and as conditions warrant, repair and replace all portions of the vertical transportation equipment included under this contract with the following exclusions only:

A. Repairs required because of negligence, accident of misuse of the equipment by anyone other than the Contractor, his employees, subcontractors, servants or agents, or other causes beyond the Contractor's control except ordinary wear.

B. Repair or replacement or building items, such as hoistway or machine room walls and floors, car enclosures, car finish floor material, hoistway entrance frames, doors and sills, telephone instrument and signal fixture faceplates, smoke detectors and communication equipment not installed by an Elevator Contractor, cleaning of car interiors and exposed portions of sills.

C. Mainline and auxiliary disconnect switches, fuses and feeders to control panels.

D. Lamps for normal car and machine room illumination.

E. Underground hydraulic piping and cylinders.

When, as a result of an examination, corrective action is found to be the responsibility of the Contractor, the Contractor shall proceed immediately to make (or cause to be made) replacements, repairs, and corrections, When such work is determined not to be the Contractor's responsibility, a written report signed by the Contractor shall be delivered to the Owner for further action.

In performing the indicated work, Contractor agrees to provide parts used by the manufacturers of the equipment for replace-

(Continued)

ment or repair, and to use lubricants obtained from and/or recommended by the manufacturer of the equipment. Equivalent parts or lubricants may be used if approved in writing by Owner.

Parts requiring repair shall be rebuilt to "as new" condition. No parts or vertical transportation equipment covered under this contract may be permanently removed from the jobsite without written approval by the owner. This does not include renewal parts stocked on the job by Contractor, which shall remain its sole property until installed for use on the equipment.

SPECIFICATIONS

9. PERFORMANCE REQUIREMENTS

Contractor agrees to maintain the original contract speed, the original performance time, the original door opening and door closing time, and the original stopping accuracy.

In accomplishing the above requirements, Contractor shall maintain a comfortable elevator ride with smooth acceleration, retardation and a soft stop. Door operation shall be quiet and positive with smooth checking at the extremes of travel. Performance requirements indicated are minimum standards, and are not the sole criteria for judging the Contractor's performance.

10. SPECIAL CONDITIONS

A. The Contractor shall post a preventive maintenance schedule and a work log in each machine room. The log shall include all entries for routine maintenance and repairs, including Supervisor's surveys. Entries shall include date work is completed, Mechanic's or Supervisor's name, brief description of work completed (including number of elevators serviced) and the approximate time required for the work. The log and maintenance schedule shall be maintained in each machine room. Owner may inspect and copy the log and maintenance schedule at any time.

B. Contractor shall maintain Owner's complete set of straight-line wiring diagrams showing "as built" conditions with any changes or modifications, to circuits resulting from control modifications, parts replacement or equipment upgrades. The Owner may reproduce these "as built" drawings and retains sole possession of these drawings in event contract is terminated.

C. State or City inspection fees shall be paid by the Contractor. Contractor shall install current operating permits in each elevator and supply the Owner with a copy of each certificate.

D. Neither this contract nor any interest therein nor claim thereunder shall be assigned or transferred by the Contractor or Owner except as expressly authorized in writing by the other party. No contract shall be made by the Contractor with any other party for furnishing any of the work or services herein contracted for without the written notice of the Owner.

E. Contractor shall assist with periodic inspection and testing of the firefighters' service in accordance with ASME/ANSI A17.1 Rule 1002. 2f and standby power operation in accordance with ASME/ANSI A17.1 Rule 1002.2g. Test will be scheduled by the Owner during regular hours two times per year.

F. If an elevator is shut down for more than 72 continuous hours (except for prescheduled or major equipment repairs). The maintenance billing for that elevator shall be suspended until the unit is restored to service.

G. Quarterly, the Contractor shall provide and review with the Owner a summary of all call-backs. The intent of this monthly summary is to minimize call-backs by keeping the Contractor and Owner aware of call-back trends.

SPECIFICATIONS

H. All maintenance or repair personnel shall sign-in with the Chief Engineer or his designee when performing work.

(Continued)

11. PREVIOUS REPRESENTATIONS

All previous communications or agreements written or verbal, are hereby abrogated and this writing constitutes the whole agreement between the parties hereto.

12. EXTENT OF LAW

This agreement shall be interpreted in accordance with the laws of the State of _____.

13. TIME
Time shall be of the essence in the performance of the terms of this agreement.

FIELD INSPECTION

This is to certify that I have this date conducted a Field/Site inspection as required by the above numbered contract.

I have contacted the person named in the Contract, or their assignee, and am satisfied with conditions as specified.

Any unforeseen conditions not specified in the contract and as found by my Field/Site Inspection are shown on the back of this form and/or attached sheets.

_____
NAME (signature)

_____
COMPANY

_____
OFFICIAL CAPACITY

_____
TELEPHONE NUMBER (AREA CODE)

Contracted Services 51

NOTE: This form must be completed and returned with Bid Proposal, or the Bid Proposal may be rejected at the discretion of the Purchasing Agent.

INSPECTION CONFIRMED BY:

_____

_____
                              Date

PROPOSAL

The undersigned declares that he has carefully examined the Advertisement for Bids, the Proposal Form, General and Special Conditions and Specifications identified as Contract Document Number _____ for INDUSTRIAL CHEMICAL COMPOUNDS FOR _____ as prepared by _____ and that he has familiarized himself with all of the conditions under which it must be carried out and understands that in making this proposal he waives all right to plead any misunderstanding regarding the same

Parts, Inspection and
Service Charges
                              (per month)
Scheduled Elevator
Maintenance $ _____ × 36= $ _____ total

Anticipated Parts and
consumables $ _____ total

Dead Weight Tests
of cabled units $ _____ total

Other parts, materials and services
not otherwise covered herein $ _____ total

                              Grand Total _____

                                        (Continued)

> CERTIFICATE OF QUALIFICATION, CORPORATION FORM. NOTARY, ETC., MUST BE FILLED OUT PROPERLY WHEN SUBMITTING THIS PROPOSAL. LETTER OF AUTHORIZATION FOR SIGNATURE MUST BE ENCLOSED WITH THIS PROPOSAL WHERE APPLICABLE. FAILURE TO DO SO SHALL SUBJECT BIDDER To DISQUALIFICATION.

Boiler Plate

Whether you decide to have the vendors input into the process in order to negotiate the best bang for your buck, or you dictate your requirements with an eye on the lowest bid, whichever document you prepare will need to be paraded past the corporate legal beagles for final scrutiny prior to its insertion into the Company Boiler Plate. How's that? Sure. The boiler plate is the tried and true instrument of commerce an organization utilizes to exact the terms and conditions surrounding your specification in order to make it "iron clad"; lawyers love 'em! In them you'll find such diverse items as:

- Participation by minorities
- Utilization of female-owned businesses
- Hiring of union trades

I don't want to bore you with the particulars here, but should you care for a more detailed view, I've included one in the Appendix for your edification. Enjoy!

CHAPTER 4

COMPUTERIZED MAINTENANCE

You're in luck! I just finished a book devoted entirely to computerizing the maintenance function. If you want to get the gory details, by all means, pick one up at your local news stand, but this "Cooks Tour" should suffice to acquaint you with the process. Since our focus is on the systems side of the physical plant equation, we'll leave the structural stuff for the Manager of Buildings and Grounds to contend with while we concentrate on setting up a program for maintaining our equipment.

The PM Function

Assuming you're already involved in some aspect of facilities management, power plant operations or building maintenance, it would be inappropriate (if not boring) for me to present you with a treatise on preventive maintenance at this juncture of your career. Rather, I'll climb up onto my soap box and expound on its mechanical virtue.

Without sounding too much like the back of a box of toothpaste, let me share with you—"Petrocelly's Axiom on the PM Process," to wit: "...preventive maintenance has been shown to be an effective, decay preventive process which can be of significant value when used in a conscientiously applied program of lubrication and regular professional care." That being the case, we naturally want to install an effective program here. Though we will consider the total maintenance picture while putting it together, remember that our primary charge is strictly to provide for scheduled (preventive) maintenance of the systems and equipment, extending and located throughout the plant.

Information Gathering

Depending on how much information was gleaned from the equipment nameplates when we first inventoried the power plant, you may or may not have sufficient data for entry into the computer, depending on the program. More than likely, you'll have to conduct a second inventory anyway, at which time you'll want to tag, number or otherwise identify each system or component, utilizing the data collection forms accompanying the maintenance management software package purchased. Nameplate data is only a small part of the information you'll need to acquire. In developing an equipment file (history card), other pertinent intelligence might include numbers, manufacturers, prices and availability of commonly replaced parts; sizes and inventory levels of consumables such as solid state boards, belts and lubricants; recommended maintenance procedures and intervals, as well as operating instructions and start up, testing and shut down procedures. (You can never know too much about mechanical devices!)

Choosing a Software Package

The likelihood is (as was the case with us), you'll purchase a software package that can do much more than just monitor equipment maintenance in your plant. Don't limit your capability for the sake of a few dollars. On the other hand, don't throw your money away on something you'll never use. But do make certain your choice meets all of your present requirements as well as future expectations. A good package will:

- possess planning and measuring tools (work prioritization and productivity reporting)
- provide for scheduling and supervising (work lists and job assignments)
- emulate existing systems and procedures (compatible with the programs from which its data is derived)

Computerized Maintenance

- have adequate financial monitors (track parts, labor and materials expenditures)

- allow managerial flexibility (access levels, customizability, etc.)

- be well organized and easy to use (menu driven and user friendly!)

Given that you've thoroughly reviewed the preceding chapter and determined how you intend to go about the search and purchase process, here are some bells and whistles you may want to include in the specification section of your bid package.

- generation of corrective/preventive maintenance work orders and schedules

- listing and updating of equipment files

- logging of work backlogs and status of equipment repairs awaiting parts

- comparisons of individual productivity to shop average or standards

- compiling of information and subsequent generation of reports

- on screen viewing and manipulation of equipment work histories

- printouts of schedules, work orders, equipment master lists and inventories

- graphic depiction of archived data such as energy usage, utilities consumption and employee productivity

- comparisons of forecasted budget figures to actual expenditures

Just a few "words to the wise"... when reviewing the various software packages available to you, make sure you're comparing apples to apples, listing all aberrations from the norm so that they can be considered separately. Disregard any obvious hyperbole and/or claims of product capability and, by all means, get any on-the-spot promises or commitments put forth by the vendor in writing. Figure 4-1 is representative of a specification one might include in a request for proposal when comparing maintenance management software packages. As always, feel free to use it as your base instrument, massaging it to meet your particular applications.

Figure 4-1.
Maintenance Management RFP Software Specification

A. GENERAL REQUIREMENTS

1. <u>Scope</u> - The software package will primarily be utilized to monitor all facets of corrective and preventive maintenance planned and performed by department personnel, track consumables and spare parts inventories and chronologically document equipment repair activity.

2. <u>Cost Allowance</u> - To be based on how well the software meets or exceeds the specifications as outlined herein and the financial implications of acquiring the hardware required for its support.

3. <u>Project Timeline</u> - The 'system' must be on line and functional prior to

4. <u>Vendor Support</u> - The software will be purchased contingent on allowing maximum input by the purchaser while requiring a total commitment of assistance on the part of the vendor through all phases of its installation and implementation.

5. <u>Training</u> - Comprehensive training must be provided for all supervisory personnel to thoroughly acquaint them with the system, such that they will be capable of adequately orienting their employees to it.

6. <u>Security</u> - System access must be limitable at more than two levels; with the lowest level incapable of data modification.

B. SYSTEM MODULES

1. <u>Interdependency</u> - All modules must be capable of operating as "stand-alone" entities as well as totally interfacing with other modules plugged into the system which access a common data base.

2. <u>Expandability</u> - The system must be able to accommodate incremental additions of new modules as they are developed and exchanges of upgraded versions must be 100% transparent.

3. <u>Alternative Capabilities</u> - All other modules available for the system outside of the dictates of this specification must be presented under a separate cover (i.e. hazardous materials, project management, asset management, graphics... etc.).

C. OPERATIONAL CHARACTERISTICS

1. <u>Work Control</u> - Software must:
   a) be menu driven
   b) create, automatically number, modify and track work orders
   c) calculate labor/material costs
   d) differentiate between maintenance types
   e) log historical data

2. <u>Materials Management</u> - Software must:
   a) be menu driven
   b) interact with work order modules
   c) provide vendor information
   d) automatically update stores
   e) flag EOQ reorder points

3. <u>Employee productivity</u> - Software must:
   a) maintain background information on employees
   b) track productive vs. non-productive time
   c) monitor overtime; worked and refused
   d) calculate mean repair/response times
   e) trace employee activity by cost-center

(Continued)

4. Documentation - Software must:
   a) maintain updated inventories
   b) provide current master equipments lists
   c) generate failure and repair histories
   d) produce work order backlogs
   e) automatically update equipment repair histories

D. HARDWARE REQUIREMENTS

Vendor will supply a complete list of all system components required to support their programs. The information should include; suggested makes, availability, discounts offered and anticipated delivery lead times.

Caveat Emptor

For those of you fortunate enough not to have been subjected to the dead language, Caveat Emptor is Latin for... "let the buyer beware!" Just as you would with any major purchase you'd make of a personal nature, say for a house or car, you need to do your homework before making a final decision on a maintenance software package.

Actually, it may pay you to do a little "extra credit" research to boot; for unlike an automobile or house which can easily be traded in or sold, once you've installed the preventive maintenance program, you'll pretty much be stuck with it. So here it is... when planning and implementing the computerization of your maintenance function, make certain to avoid:

- buying without comparing product capabilities and prices (*caveat emptor*)

- CPU's and peripherals having slow operating speeds (unnecessary downtime and employee tie ups)

- selecting operating systems incapable of future expansion (initial set up costs versus new set up costs)

- purchasing software that doesn't fit your operating needs (don't use a wrench to pound a nail)

- buying hardware before deciding on the software to be used (like chartering a bus before you know the number of riders to be transported)

- skimping on memory capacity (storage and retrieval of information is a computer's main function)

- inadequate vendor maintenance and support coverage (you can run it, but can you fix it?)

- multi-user computer set-ups (unwanted access to information by others and slowing of the system)

- unreliable or untested software/hardware (bugs will drive you buggy—inaccurate data)

- errors in estimating needs (if you don't double check your figures it can cost you twice as much)

- buying on "the cheap" (you get what you pay for)

- unrealistic expectations of program and operator abilities (GIGO - garbage in equals garbage out)

- taking on too much too soon (you've got to learn how to walk before you can run)

- pressuring employees to learn and use the computers too adamantly (you can lead a horse to water...)

Figure 4-2.
Computerized Maintenance Management Software Comparison

Focus:	A	B	C	D
Equipment Entries-10,000	Yes	Yes	Yes	Yes
Key Control System	Yes	No	No	No
In Service Record	Upgrade	No	No	No
Master Equipment Listing	Yes	Yes	Yes	Yes
Daily Rounds Procedures	Yes	No	Yes	No
Equipment History	Yes	Yes	Yes	Yes
Archive Feature-Selective	Yes	Tape	Tape	Tape
Word Processing	No	No	No	No
Support 386, 33MHZ, Laser	Yes	No	Yes	No
Part Inventory Package	Yes	No	Yes	Yes
Recorder	Yes	No	Yes	Yes
Pricing & Vendor File	Yes	No	Yes	Yes
Integrated Purchase Order	Yes	No	Yes	Yes
Energy & Graphics Usage	Yes	No	Yes	No
Job Instruction	Yes	Yes	Yes	Yes
Engineering Data	Yes	Yes	Yes	Yes
Operating Specifications	Yes	Yes	Yes	Yes
Corrective Work Orders	Yes	Yes	Unclear	Unclear
Sequential Number	Yes	Yes	Yes	Yes
Status Codes	Yes	Yes	Yes	Yes
Work Done Codes	Yes	Yes	Yes	Yes
Customized Format	Yes	Yes	Yes	Yes
Time Standards Work Load	Self Loaded	Pre Loaded	Self Loaded	Self Loaded
Dpt. Codes for Charge	Yes	Yes	Yes	Yes
Warranty Exp. Tracking	Yes	Yes	Yes	Yes
Outside Contract Coverage	Yes	Yes	Yes	Yes
1 yr. Technical Support	2 years	Yes	Yes	Yes
Label Making	Yes	No	No	No
Password Protection	Yes	Yes	Yes	Yes
Small Project Mgmt	Limited	No	Yes	No
Under $8000	Yes	Yes	No	Yes
Network Capable	Yes	Yes	Yes	Yes
Customizing Options	Yes	Yes	No	Yes
Employee/Trades	Yes	Yes	Yes	Yes
Priority work order schedule	Yes	Yes	Yes	Unclear
Relational Database	Yes	Yes	Unclear	Yes
Quality Assurance Tracking	Yes	Yes	Yes	Yes
Room Inventory & Specs	Yes	No	Yes	No

Once you've received the bids back and had an opportunity to discuss the relevant features and benefits of a few software programs (and review a demo disk or two?) it's time to subject the proposals to closer scrutiny. As you're going to have to live with this thing after its installed. So remember to check the vendors offerings against your original requirements and...

- assure yourself they are providing for the applications you requested

- there are no hidden costs over and above those you are tentatively agreeing to

- the hardware platform has been recommended and is 100% compatible with the software intended for it

- the programs are easy to understand and learn

- check on system history and reliability claims through referenced users

- get commitments regarding updates, training, warranties and service agreements

- arrange for real life demonstrations of the hardware and software in concert

- lock in prices, promises and refund/exchange policies

- make sure appropriate insurances are in place prior to start-up

Installing the Program

If you avoided the pitfalls normally associated with buying the software, I'll concede that you've been assisted with the purchase of the proper hardware and that you're reasonably familiar with its operation and peripherals. That so? Great! Then what's needed now is your undivided attention for as long as it takes for you to gain an understanding of what you'll need to do to bring it all together.

The next step is the most critical in the installation process. One that, if ignored, can put the kibosh on your whole program. If you don't understand how a thing works, you can't make it run. READ THE MANUAL! I don't care if the paint is peeling off the walls, last week's change orders weren't approved or the unions are ready to walk out; delegate the problems to someone else and... READ THE MANUAL! Just as the most important aspect of real estate business is location, location, location; understanding, understanding, understanding is of paramount significance in the computerized maintenance game. READ THE MANUAL!

Now that you've devoted the necessary time to thoroughly acquaint yourself with the software and you've reproduced adequate numbers of forms and collected the data you'll feed into the computer, the remainder of the process will be a simple, if not mundane activity. I suggest you leave the keypunching to those with the digital dexterity to accomplish the data entry function. Once entered, you can then manipulate the information to your heart's content while you determine inspection frequencies, add and delete maintenance procedures and the like. Happy computing!

Chapter 5

Manpower Requirements

We probably shouldn't put the staffing thing off any longer. You know what shape your plant is in, which services you plan to contract out and how many preventive maintenance man-hours your equipment will require annually. Before deciding on the number of people you'll need to run the place, you should first determine your department's goals, evaluate the existing staff's capacity for achieving them, then actuate a plan to make up the difference. In this way, you'll not only ascertain the proper staffing levels but the appropriate skill sets, as well. This is an opportune time to establish a core of expertise on which to build your future operations. Here are a few bricks to set on your foundation...

The Screening Process

As plant managers, the quality of our operations is judged on the basis of our employees professionalism, performance and public relations ability. In short, we are only as good as our people. Now I know that adage is old wisdom, but it's wisdom nonetheless. If we surround ourselves with competent personnel who always get the job done, properly and on time, we will be perceived as competent ourselves. If our personnel lack the proper skills or initiative, we will be considered poor leaders and incapable of managing. None of us is fully blessed with the best or cursed with the worst. We are in a position, however, to mold our people to fit into either group. How successful we are in educating and motivating them is the true test of our management ability.

Most companies subscribe to preordained hiring practices by policy. Their standard employment applications are not department or job specific since they are used throughout the company, and pre-employment testing of applicants by the personnel department doesn't always work. How do you find the right people to fill all the different positions in your department?

If you're an expert in every trade, that won't be a problem for you, but you're not. None of us is. I know it's hard to admit, but come down from your ivory tower; none of us is expected to be. Would you feel comfortable interviewing journeymen mechanics and tradesmen to fill all of your department's positions? If you can honestly answer yes to that question you don't need my help or anyone else's for that matter. But if you're like me, you can use all the help you can get. So how do we go about it?

After explaining the minimum requirements you'll accept, your company Human Resources Department can save you a lot of time and grief by thoroughly screening applicants for you. They can research their work backgrounds, check personal and business references, determine academic achievement and summarize any special skills they may possess. As the result of screening, you might then be confronted with 10 candidates instead of 400 applicants. The requirement levels you set determine how many candidates you will see. Establish realistic parameters based on available manpower from the community from which you expect to draw the candidates. If you set a goal too high to attain, you may only be talking to yourself. Remember, you can train them after they're hired. Ultimately you'll interview every person who will eventually become a part of your organization, but should you be the one to determine their qualifications? Let your supervisor further screen the candidates for you. If your lieutenant thinks they're qualified, then they probably are. Your job will be to determine how well the candidates will fit in with the rest of your group.

After the candidates have passed muster with your technical supervisors, by all means give them your personal once-over. Keep the proceedings on an informal basis and don't resort to trick questions. Be honest and try not to psychoanalyze every answer you get. Keep it simple and to the point. It hasn't been so long since you were in the hot seat on the other side of the desk. Remember how you felt and give the guy a break. Share a job description with him (her) and discuss each duty listed. Have him discuss similar duties he has performed for other employers. Does he appear knowledgeable on the subject? Post a hypothetical problem and ask him how he would resolve it. Was his response logical and authoritative?

After the interview, the screening process continues. *You* may want to discuss the candidate's qualifications with your supervisors or even call them back for additional interviews. Your company may require pre-employment physicals. After hiring them, the probationary period serves as a screening tool in determining if they will remain on as employees. Even if they make it through their probationary periods, they must pass the scrutiny of their supervisors daily and do well on their annual performance appraisals. A person may be hired as a full-time employee but there is no such thing as a permanent position.

Position Descriptions

No job description has ever been written that completely and totally defines every element of every duty required to be performed by an employee working in a power plant, regardless of his/her station. Why have them? Because, properly utilized, they are an excellent tool for communicating a manager's expectation of subordinate performance levels and a good gauge against which efforts can be measured and personal development assessed. Therefore, a reasonable amount of thought should go into their construction.

Job descriptions are actually position descriptions. They document the responsibility of titled positions. They can be comprised of simple lists of duties to be performed by individuals in the course of their work, ornate sets of instructions detailing the smallest aspects of a position or anything in between. The idea, then, is to avoid writing them as though they were work assignment sheets and, instead, capture the essence of the work you're describing and will ultimately communicate to the person(s) they are being written for. The base of understanding they provide goes a long way in establishing the department's direction and sorting out misunderstandings on the job.

Every position and level within a position warrants that a separate description be written for each. As no job description, however comprehensive, can anticipate every conceivable future circumstance, it's always wise to include a statement indicating that the duties in its text are a summary of the responsibilities associated with the position and are not intended to address every task the holder may be called upon to complete. A notation such as this can belay many arguments you may otherwise encounter should your employee try to use his job description as a defense against inferior performance. Let me share a PD with you that I've recently created for a Director of Facilities Management position.

Figure 5-1. Job Description

Job Description

Job Title: Director of Facilities Management

Summary: This position is responsible for administering and directing the management of the Plant Operations, Facilities Maintenance, Safety Management, Security, Communications and Engineering Departments. It monitors the planning and organization of and exercises control over all aspects of facilities management throughout including design, construction, alteration, purchase, installation, operation, maintenance, and disposal of systems, structures and equipment. It is also responsible for overseeing loss prevention functions including safety, security and fire protection: Telecommunications management, including telephone, paging, television and radio communication: equipment service and repair, as well as administrative functions such as establishing objectives, standards of performance, policy and standard operating procedures and monitoring regulatory code compliance.

Qualifications:

- Bachelor of Science (B.S.) Degree in a relevant engineering discipline

- Professional Engineer (P.E.) or Certified Plant Engineer (C.P.E.) designation

- Certification as a Property Manager (F.M.A., C.P.M.)

- 10+ years progressive experience as a Facilities Manager in a comparable organizational setting

- Prefer - Certified Engineer (.) and /or Chief Engineer licensed background and experience

Knowledge/Skills Required:

- Comprehensive understanding of federal, state and local codes and standards

- Experience in dealing with regulatory and accrediting agencies.

- Man-hour, operating and capital budget preparation

- Construction project planning and management

- Background experience in the trades and/or power plant

Principal Responsibilities:

- Consult with administration and fiscal services concerning department budget, major purchase priorities, major construction projects, renovation, and similar projects requiring large capital expenditures. Prepare annual capital and expense budgets and ensure all departments operate within allocated funds. Monitor annual capital equipment budget for disposal and disposition of equipment replacements. Develop annual facilities project list. Evaluate and approve or disapprove equipment acquisitions included in annual capital equipment program prior to procurement by purchasing department. Maintain appropriate records and statistics.

- Develop and implement departmental goals, plans and objectives. Assist Directors in organizing their departments and developing methods for improving their operations. Ap-

(Continued)

prove establishment and revision of department policies/procedures. Annually evaluate the performance of the Department Directors under his/her span of control.

- Provide consultation and guidance to key personnel. Interpret hospital and department policies and procedures, standards, requirements and regulations of accrediting and regulatory agencies, and national, state and local electrical, building, and plumbing codes to department staff Update practices, policies, and procedures as needed to remain in compliance. Review department practices, quality of work performed, production schedules, and effect changes as needed to improve services. Regularly inspect buildings and grounds to assure conformance with established standards and regulations. Establish reporting and approval procedures and delegate to Directors sufficient authority to direct the work of the employees reporting to them, as to selection, orientation, training, job objectives, performance standards, methods, performance appraisal, personal development, transfers, promotions, terminations, vacation schedules, control of hours approval of time off, time recording and payroll authorization, salary adjustments, safety communications, discipline, settlement of grievances and other personnel actions.

- Conduct determination of need and cost benefit studies for proposed new programs, equipment and services.

- Establish standards of performance and productivity; establish a preventive maintenance program; establish quality assurance and safety programs in cooperation with quality assurance and risk management personnel.

- Determine staffing needs: interview, select, orient, train, and supervise department directors, work methods and performance standards. Direct their work and appraise their performance relative to department objectives. Initiate, authorize, and communicate promotions, transfers, discipline, discharges, vacation schedules, time off and all other personnel actions.

- Perform ancillary duties such as membership on hospital committees and acting as the liaison in dealings with architects, engineers, contractors, code administrators, public utility representatives municipal authorities and other outside professionals and agencies. Coordinate with departments outside his/her authority in developing space plans for new construction or repairs and space utilization programs. This includes preparing engineering maintenance cost estimates for renovation work, architectural drawings, and developing construction timetables.

- Plan and recommend development of physical facilities. Review and recommend approval of plans for construction. Advise on structural changes and additions/modifications to buildings. Interview contractors to receive and analyze bids, basing recommendations on economy and feasibility of bids.

- Direct and may perform purchasing functions of the department. Evaluate and justify department needs and purchases. May meet with sales representatives, negotiate contracts, bring in outside trades and technical specialists, arrange for all purchasing of maintenance department supplies and equipment and perform related functions. Direct receiving, storage, and inventory control functions.

- Execute professional consulting, engineering, construction, alteration, preventive maintenance, operation, material, and equipment contracts within authorized resources. Administer construction contracts and monitor construction in progress to ensure compliance with plans and specifications. Investigate and resolve complaints relative to noise, dirt, and continuance of services while projects are in progress. Interface with environmental services to assure custodial services. Coordinate activities with local and state jurisdictions in obtaining certificates for occupancy. May negotiate real estate transactions and interface with real estate brokers as necessary.

(Continued)

- Plan for major repairs, renovation, new construction. Investigate needs for major repairs in collaboration with administration and department heads. Periodically inspect buildings, grounds, and power plant, evaluate condition and needs and take or recommend corrective measures. Work with department heads in developing floor plans for new or revised areas. Interface with and represent administration with professional consultants in areas of facilities planning, engineering, and management. Make recommendations on major repair and construction projects as to work to be done by outside contractors or by personnel. Analyze quotes, make recommendations for acceptance, and serve as professional liaison between and contractor during construction cycle.

- Direct plant engineering capability. Develop systems for control and efficient and effective utilization of all utilities such as gas, water, and power. Confer with utility companies, city and state inspectors, and insurance companies regarding functional activities. Participate in insurance inspections and claims.

- Promote and encourage professional growth and development of personnel through attendance at in services, workshops, seminars, etc. Review accident reports and take appropriate action. Promote intra and interdepartmental cooperation. Participate in union negotiations, administer labor agreement, and settle grievances.

- Approve/Review and cause to be prepared, various periodic reports to administration which include but are not limited to; employee overtime reports, monthly activity reports, biweekly boiler efficiency reports, energy utilization reports and orders for space, material, equipment and services necessary to carry out department functions.

- Maintain department records, reports, statistics, files as required and perform related duties. Establish a record keeping system, repair and maintenance schedules, charge schedules, forms used by department. Prepare project reports and main-

> tain historical data as required. Establish current files of as-built, schematic, and engineering drawings of plant facilities and equipment, including specifications, operation and maintenance manuals.
>
> - Participate in research activities. Study and evaluate new products, services, and technical equipment in order to assess their value in improving productivity in the department.
>
> - Maintain an adequate reference library of manuals and texts accessible to department personnel.

Yours needn't be this elaborate. This particular description was constructed such that it would attract only a narrow group of individuals (abilities) into contention for the position, assuring the hiring of a properly qualified person in filling it. Too often, PD's aren't specific enough about ability and/or experience requirements, leading to improper placement of persons lacking the appropriate skills needed for the job. This holds true for all levels and positions in the power plant.

Figure 5-2

> PLANT ENGINEERING JOB DESCRIPTIONS
>
> CHIEF ENGINEER
> Supervises the operation and maintenance of steam boilers and engines, turbines, vacuum and centrifugal pumps, heating and ventilating systems, refrigeration equipment and machinery required for building operation and maintenance.
>
> SUPERVISES EQUIPMENT OPERATION:
> Plans and schedules work for the operation of all types of equipment in the various buildings and locations. Sets up staffing schedules for all shifts and plans for relief and emergency staffing.
>
> *(Continued)*

Gives supervisory assignments covering daily operations to Operating Engineers II and Operating Engineers I, and gives instructions on the testing schedule for all machinery, fire fighting equipment and safety devices, as well as instructing supervisory personnel in procedures and planning for conduct of preventive maintenance. Follow up on the satisfactory completion of assignments either by personal observation or through progress reports.

Prepares requisitions for new equipment and repair to existing equipment by studying what is needed and how best to secure the most effective results. Originates requests for modernization and improvements to the facility, makes recommendation by survey of existing conditions and means to provide ever increasing service. Supervises fire brigade in absence of fire marshal.

Responsible for the general comfort and safety of a large number of employees in relation to heating and operating services at the visitors, main and other buildings on the complex. Is direct supervisor to heating and operating employees; such as Operating Engineer II's, Operating Engineers I, Firemen, Boiler Washer, Mechanical Assistants and various tradesmen. Prepares operating and personnel reports for management purposes. Trains, supervises and disciplines various employees under his control, in various locations of the   complex.

ASSISTANT CHIEF ENGINEER
Works with the Chief Engineer to see that all orders of the Chief are carried out. Supervises work which is done in the powerhouse and related facilities. Assigns and schedules work to be done in order to efficiently run a powerhouse. Performs administrative duties, such as employees time.

Has full responsibility in maintaining discipline in absence of Chief Engineer. Periodically supervises inventory of stock. Creates programs, schedules and instructions for employees. Assigns all job maintenance in accordance with the instructions of the Chief Engineer.

Assigns a working diary of all repairs, replacements and new installations in the powerhouse.

Makes daily inspection tours of entire complex. During these inspections, routinely checks the boilers, generators, high voltage vaults, switch boards, pumps, valves, control equipment, water, steam and electrical services. After inspection, submits a report to the Chief Engineer. Checks inventory of tools, spare parts, fuel oil, operating and maintenance supplies for the powerhouse. Logs, charts, meters and gauges for any deviations from normal operation are also checked. Supervises the cleaning, repairing and testing of valves, air conditioning and auxiliary equipment.

BOILER ROOM AND SHIFT ENGINEERS
Duties require that the engineer stand watch in the powerhouse. Under no circumstances does the engineer leave the powerhouse unattended. General supervision of all shift personnel. Responsible for the operation of the powerhouse switchgear, transformer, 2-800 ton central air conditioning units including cooling towers, fans, pumps, and all auxiliaries, also fire pump operation.

Supervises boiler room personnel and 3 high pressure boilers fired by oil and gas. Between the hours of 7:00 a.m. - 7:00 a.m. (24) twenty four hours a day on all holidays and weekends, an engineer is in charge of the powerhouse and physical plant operations. These include all emergency and routine calls, emergency repairs, elevator emergencies, fire calls, and all other "trade" emergency calls that can be handled locally while the trades are not present.

Maintains daily logs for the engine room and other logs; such as emergency and routine repair calls, elevator log for contractor arrival and departure, logs calls for trades due to emergency work and pick up for waste management.

BUILDING ENGINEERS
Under supervision of shift supervisor, makes a complete tour of the complex every two hours, which is to include; domestic water pumps, vacuum pumps, condensate pumps, compressors, fire pumps, pneumatic system, fan rooms, and air units. This tour also includes 2 emergency generators. Answers all routine and emergency calls, requiring immediate attention. All calls requiring

(Continued)

service the next day, must be logged. Answers all fire calls. Assists supervisor engineer as required for start up and shut down of central air conditioning equipment and auxiliaries. As required assist in boiler room operations such as, change over of boiler and auxiliaries.

SHIFT FIREMEN

Under supervision of the shift engineer as required in the general operation of the boilers and auxiliary boiler equipment. Changes burners and makes adjustments. Logs information related to the boiler hourly. Maintains proper blowdowns on the boiler and will undertake a soot blowdown if necessary. Is responsible for the general upkeep and cleanliness of boiler room, boiler equipment and machinery as well as cleaning of associated rooms throughout the complex and work area. Makes a routine inspection of boiler room equipment and machinery. Inspects the boiler water treatment facility, boiler fans, pumps, and other equipment not necessarily located in the boiler room. Performs duties as required by the shift supervisor and shift engineer. Work orders issued daily from the Chief Engineer to be completed while on duty or show cause.

MAINTENANCE CREW

Responds to all heating and air conditioning calls. Must calibrate or repair as required all pneumatic temperature and humidity controls in all areas, including patient and personnel care areas throughout the complex. Inspects, maintains and makes adjustments or repairs as required on all air handling units or areas with sophisticated controls on a daily basis. Change filters, roll-omats, dry packs. Change filters in the mechanical rooms. Assist trades in repair or replacement of any engineering equipment. Grease and/or change oil in all air handling units, vacuum pumps, compressors, domestic water pumps and sump pumps. Change V-belts as required, clean oil and/or grease all portable fans. Clean all control stats and exhaust equipment. Clean and mop all machinery and other engineering areas. Make all utility shut downs for repairs or replacement as needed. Inspect and adjust as required all hot water heaters. Make all fire calls and do preventive maintenance cards as issued on preventive maintenance program by the Chief Engineer.

Shortly after arriving at this facility, I asked the Chief Engineer to give me a copy of the job descriptions for all of the power plant positions. Figure 5-2 was what I received. Though it accurately summarized the department's activities, this "one" document was far from the individual position descriptions which I was expecting (hoping for?). There was no effective date or mention of who authored it, no review or approvals, no required qualifications (other than implied by the license level held), no signatures, etc. And all the jobs were lumped together. Looks like we have our work cut out for us. As an aid in developing job descriptions for your staff, I've included a list of mechanic related job duties/traits in the Appendix for your edification.

Licenses and Permits

For those of you who are following my literary efforts, the last time I delved into this subject I was in New York. Well I'm in Chicago now and... what a difference a state makes! Checking the city's statutes regarding licensure of operators has really opened my eyes. Without delving into the magnitude of the dissimilarities between the two, let me suffice it to say that, nationally, licensing of power plant operators is "all over the board!" As it would be impossible to hit everyone's nail on the head in this regard, I'll just hammer the point home that it would behoove you to acquaint yourself with the rules and regulations of the governing authorities under whose jurisdiction you fall.

Power plants are permit and license intensive facets of all organizations, the operations of which fall under the scrutiny of innumerable government departments, regulatory bodies, accrediting agencies and trade associations. Some have the authority to fine and imprison you, some to shut you down, and others just to soil your reputation, but all of them can cause you to soliloquize for hours on end. Rather than list all the possible laws you may be about to break, I thought it might be better that I give you some "rules of thumb" to follow in order to avoid the heartache (and heartburn!).

1. Find out under whose auspices you fall and maintain an up-to-date set of their codes (summaries, subscriptions) in your office.

2. Make sure that every piece of equipment requiring an operating certificate has one. Post a signed copy of the certificate by the equipment and keep the original in the office files.

3. Schedule equipment shut downs for certificate inspections well in advance to avert unanticipated repairs or conflicts with the inspectors itinerary.

4. If certificates are due or past due, don't wait for the inspectors to contact you—put them on notice (preferably in writing) that you were (are) prepared for the inspection and have them send you correspondence on any deferments caused by them.

5. Never operate equipment lacking the proper certification nor allow equipment to be operated by persons not properly licensed.

6. Make certain all equipment and operators licenses are current, posted and on file in your office and that all license/certificate invoices are paid promptly upon receipt.

Staffing Levels

As was stipulated earlier, we aren't going to concern ourselves with the building trade's side of the operation, only the power plant's. But it must be mentioned that (in our case) the operating engineers will be called upon (outside of normal business hours) to provide emergency maintenance to the facility in the absence of regularly scheduled tradesmen.

The following is a staffing scenario relative to the operating needs of the power plant for monitoring incoming utilities, generating steam and chilled water for comfort/process, applying the preventive maintenance program, testing/troubleshooting/repairing equipment and maintaining the machinery rooms. I've based the staffing requirements on mandatory hours of coverage in the boiler room, those anticipated in providing for the preventive maintenance program, allowance for non-productive employee time and consideration of general operational needs; to wit,

Manpower Requirements

Figure 5-3

Requirement/Annual Hours		Equipment FTE's/License
Supervision	/4160	1.00/III
		1.00/II
Mandatory coverage of boiler room (3 shifts)	/8760	4.21/I
Relief coverage for boiler room personnel (non-productive time)	/1448.24	.69/I
Daylight & Afternoon (mechanical assistance)	/8320	4.00/Fireman
Non-productive relief of mechanical assistance	/0000	0.00/Fireman
Preventive Maintenance mechanic (as scheduled)	/14560	7.00/I
Non-productive relief of preventive maintenance mechanic	/0000	0.00/I

SUMMARY

Title	Number
Stationary Engineer Class III	1
Stationary Engineer Class II	1
Stationary Engineer Class I	12
Fireman	4

*Note: The Operating Engineers will be responsible for handling emergency maintenance calls within the hospital during off shifts, weekends and holidays.

Figure 5-4

BI-WEEKLY SCHEDULE

Pay Period Start _____
Pay Period End _____

Engineers Date _____

	Sun	Mon	Tue	Wed	Thu	Fri	Sat	Sun	Mon	Tue	Wed	Thu	Fri	Sat
Shift Operator	off	7-3	7-3	7-3	7-3	7-3	off	off	7-3	7-3	7-3	7-3	7-3	off
Shift Operator	3-11	off	off	3-11	3-11	3-11	3-11	3-11	off	off	3-11	3-11	3-11	3-11
Shift Operator	11-7	11-7	11-7	off	off	11-7	11-7	11-7	11-7	11-7	off	off	11-7	11-7
Relief Operator	7-3	3-11	3-11	11-7	11-7	off	off	7-3	3-11	3-11	11-7	11-7	off	off
PM Mechanic	off	off	7-3	7-3	7-3	7-3	7-3	off	off	7-3	7-3	7-3	7-3	7-3
PM Mechanic	off	3-11	3-11	3-11	3-11	3-11	3-11	off	off	3-11	3-11	3-11	3-11	3-11
Fireman	off	7-3	7-3	7-3	7-3	7-3	off	off	7-3	7-3	7-3	7-3	7-3	off
Fireman	7-3	3-11	3-11	3-11	3-11	off	7-3	7-3	3-11	3-11	off	off	off	7-3
Utility Man	11-7	11-7	11-7	11-7	off	off	7-3	11-7	11-7	11-7	11-7	off	off	7-3

Orientation and Training

This is actually the second writing of this section of the book. Unfortunately, after a near year on the road, I inadvertently "tossed" the original in my rush to return home. In retrospect, it could probably have been considered a superlative, definitive treatment of the subject, but unless someone rummages through my trash and sends it to a literary society... well, I guess we'll never know! And since I'm reviewing this work at the eleventh hour for "tardy" submission to my publisher; please accept my apologies, but it's hard to be creative under such circumstances. Consequently, you won't receive the benefit of my wit and wisdom, but I feel obligated to give you something—how about a couple of policies to help you in setting up your program? Great! I know you can use them. Thanks for your understanding.

Figure 5-5

PLANT MANAGEMENT		
POLICY TITLE: New Employee Orientation		POLICY NUMBER: 3-06.4.1.0
DATE OF ORIGIN: 11/6/92 REVIEWED & REVISED	FOR DIVISION USE FOR DEPARTMENT USE X	PAGE 1 OF 3
I. PURPOSE: To provide a planned introduction and orientation program for new employees to learn about their department and what is expected of them as one of its members. II. ORIGINATOR: Director of Facilities Management III. SCOPE: Plant Management staff, Plant Operations and Facilities Maintenance personnel.		
APPROVED: (Signature) (Title) (Date)		

(Figure 5-5 continued)

Policy #3-06.4.1.0

IV. TEXT

A. <u>Orientation</u> <u>Program</u>
All new employees will attend the hospitals structured orientation program at its earliest convening, in accordance with established hospital policy.

B. <u>Building</u> <u>Tour</u>
Concomitant with the in-service, new employees will be required to become acquainted with the facilities building structures and systems. Escorted by one of the departments supervisors and encouraged to ask questions, they will:

- physically traverse the facility, floor by floor, beginning on the roof and finishing in the basement

- tour the power house and all the mechanical spaces, tracing out each of the buildings systems

- walk the exterior of the main building perimeter and visit all the facilities outbuildings

C. <u>Department</u> <u>Operations</u>
Each new employee will pay a visit to the departments office where he/she will be:

- introduced to the Plant Manager and office staff

- provided with an overview of the departments scope and function

- given ample time to review all department policy manuals in a quiet surrounding

D. <u>Supervisors Orientation</u>

After completing the building tour and visiting the office, the new employee will spend time with his/her immediate supervisor with whom they will cover:

- personnel matters not already addressed by the Human Resource Department or the employee handbook

- payroll matters not already address such as time clock locations, time card usage, tardiness and absenteeism

- employee parking, ID badges, break and lunch times

- forms used for work processing and assignment

- the employees job description and departmental rounds and routines

- emergency procedures and use of fire fighting equipment

- MSDS sheets and the Employees Right to Know law

E. <u>Work Area</u>

Prior to beginning work, each new employee will be taken to the shop or area that they will work out of where:

- the shop rules and hours of operation will be explained

- they will be provided with the tools and materials required to perform their individual jobs

- they will be introduced to their co-workers

Figure 5-6

FACILITIES MANAGEMENT		
POLICY TITLE: Department Training Program	POLICY NUMBER: 3-06.4.2.0	
DATE OF ORIGIN: REVIEWED & REVISED: 2/08/93	FOR DIVISION USE X FOR DEPARTMENT USE	PAGE 1 OF 2
I. PURPOSE: It is the intent of the Facilities Manager to foster an environment of internal training and orientation which will assure that all employees under his direction are thoroughly acquainted with all aspects of their department operations and are provided with the knowledge needed for them to properly perform their work. II. ORIGINATOR: Director of Facilities Management III. SCOPE: Plant Maintenance, Facilities Maintenance, In-service Education		
APPROVED:		
(SIGNATURE) (TITLE) (DATE)		

(Figure 5-6 continued)

Policy # 3-06.4.2.0

IV. <u>TEXT</u>

The training program encompasses the whole of each departments educational efforts. It is tied in with employee orientation in-services, licenses requirements, personnel documentation, department training and employee development; thereby comprised of two component parts - voluntary and involuntary, as follows:

<u>Voluntary Component Elements</u>
- trade specific presentation (in-house)*
- vendor demonstrations (outside)*
- manufacturers service schools*
- interdepartmental orientation*
- annual reorientation (operations)*

*(Unless recommended by immediate supervisor)

<u>Involuntary Component Elements</u>
- new employee orientation*
- new employee orientation (department)*
- mandatory in-services*
- training required for license renewal
- monthly department meetings*
- external developmental training

*(Per established policy)

A record of attendance will be kept of all group sessions; a copy of which is to be inserted in SOP volume #4. Records of individual attendance and/or copies of certificates of completion and licenses will be inserted in the individuals personnel jackets.

Figure 5-7

FACILITIES MANAGEMENT		
POLICY TITLE: Trade Specific Programs	POLICY NUMBER: 3-06.4.3.0	
|DATE OF ORIGIN: |REVIEWED & REVISED: 2/08/93	FOR DIVISION USE FOR DEPARTMENT USE X	PAGE 1 OF 4
I. PURPOSE: It is the intent of the Plant Operations and Facilities Maintenance Departments to construct and implement educational programs involving all of its members for the purpose of elevating their individual skill levels, improve productivity and enhance professional image. II. ORIGINATOR: Director of Facilities Management III. SCOPE: Plant Maintenance, Facilities Maintenance, In-service Education		
APPROVED:		
(SIGNATURE) (TITLE) (DATE)		

(Figure 5-7 continued)

Policy # 3-06.4.3.0

IV. <u>TEXT</u>

These programs are designed to serve the individual training needs of all department members. Each serves as a "trade specific" educational core incorporated into the master department training and orientation program.

<u>Composition</u>
The main focus of each program is its four year (real time) progressive curriculum, similar to a standard apprenticeship program around which individual training efforts can be structured.

It's evaluation criteria delineates specific duties to be performed in each position, at the appropriate responsibility level and can be used as a checklist for assessing the competency of existing personnel for determining at which level they need to enter the program.

<u>Program Elements:</u>
- establishment of a technical reference library
- scheduled classroom space for 12-20 people
- voluntary employee participation
- paid classroom time (3 hrs/man/week)
- on the job cross training between mechanics
- one-on-one training by supervisor
- employee attendance at vendors service schools
- pre-approved, sponsored off-site instruction on employee time

(Figure 5-7 continued)

- funds allocated for audio-visual materials, textbooks and supplies

- incorporation of job descriptions

- pre qualification of existing personnel

- utilization of correspondence courses with in-house examination proctoring enabling certification of efforts

- instruction towards licensure

- periodical review of individual progress

- recording of achievements

Program Core Overview

(REPRESENTATIVE/TO BE EXPANDED)

		Approx. Hours
1.	Orientation	500
	A. Introduction to Training Program	
	B. Terminology	
	C. Maintenance Procedures	
	D. Safety Training and Fire Prevention	
	E. Codes/Licenses	
2.	Care and Use of Tools and Equipment	600
	A. Hand Tools	
	B. Power Tools	
	C. Electrical Test Equipment	
	D. Mechanical Test Equipment	
3.	Testing, Inspection, and Repair of Equipment	1500
	A. Trouble Shooting	
	B. Unit Replacement	
	C. Component Repair	
	D. Control Circuits	

4. Operations and Maintenance   1400
   A. Maintain Permanent Records
   B. Warranty/Guarantees
   C. Equipment Inventory
   D. Spare Parts
   E. Requisition for New & Replacement Parts

5. General   4000
   A. Equipment/Systems
   B. Appliances
   C. Special Procedures
   D. Operating Contingencies
   E. Safety Elements (Advanced)
   F. Emergency General Maintenance
   G. Miscellaneous
   H. Housekeeping

Total Hours   6000

Evaluation Criteria
A weighted checklist of principle accountabilities will be assembled for each position for use in pre qualifying individuals to the program and determining student progress.

Chapter 6
Boiler Basics

Ask the average Joe (or Joanne for that matter) what a boiler room is and they'll tell you it's a place where illegal commerce is conducted. To many, boiler room operations conjure up visions of quick-change artists and slight of hand. Such aspersions and misconceptions often take root in the uninitiated as the result of fear and misunderstanding. The truth is boiler rooms house some of the most gregarious, honest, hard-working and dedicated people in the work force. But even we can find ourselves in hot water (excuse the pun) if we don't take care of business down here. Such as what, you ask? How about...

Their Support Systems

Every (so-called) power plant "expert" that's come down the pike has arrived with his or her personalized definition of what a boiler is, running the gamut of simple to absurd. Assuming you have at least a nodding acquaintance with the device, if not a license to operate it, I'll spare you the usual litany normal at this juncture and suffice it to say that a boiler is "... a heat exchanger, the efficiency of which is contingent on the dependability of its operating systems." Those five separate, though interdependent systems, include piping and components for delivering make-up water to the boiler and returning spent steam to the boiler room (feedwater/condensate system); combustion air (draft system) and combustibles (fuel system) to the furnace; discharging and apportioning steam at various pressures/temperatures to the loads (steam distribution system) and providing for safe, efficient and cost-effective equipment operation (safety/control system).

As a steam boiler exchange medium is water, the *Feedwater/Condensate System* encompasses the whole of the water-steam-water cycle. The process begins when raw water is introduced into a cold boiler, where it is heated until transformed into steam and discharged under its own pressure through steam pipes to a load which extracts the

Figure 6-1

added heat, reverting the steam to condensate which is eventually reintroduced into the boiler for another trip through the loop (save for system losses). Components within this system include the steam traps which separate the condensate from the steam, the tanks and pumps which store and deliver the feedwater for and to the boiler, the controls which actuate said delivery and the chemical treatment transfer pumps.

The *Draft System* includes all components within the way of travel of the combustion air from where it enters the boiler room through the intake air louvres, through the forced and/or induced draft fans and furnace area, past the heat exchange surfaces, through the boiler breeching and up the stack.

The *Fuel System* interacts with the draft, feedwater and steam systems through a series of safety/control system interlocks. Components of this system include the nozzles used to admit the fuel into the furnace, the devices used for heating, pressurization and delivery, the vessels/piping in which the fuel is stored, interconnections for an alternate source and the programmer that determines the fuels consumption.

The *Steam Distribution System,* has at its base, the generating tubes and steam space in the boiler proper and all other valves, pipes, etc. which contain or come into contact with the vapor, including pressure reducing stations, heating coils, actuating devices, etc.

The *Safety/Control System* incorporates all the instruments/accessories that monitor, control, modulate, operate and if necessary shut down boiler operations for safety reasons. Items associated with this system include flame-eyes, low-water fuel cut-offs, feed water regulators, high/low water alarms, safety valves, etc.

Internal and External Inspections

Whether or not you are required to by some governing body or an insurance carrier and regardless if your boilers are rated as high or low pressure, its a good idea to unbutton them once a year and peak inside. Aside from the sins your operators may have committed upon them, the nature of their existence and operation alone dictates an occasional look see. They are exposed to frequent extremes of hot and cold, fed megadoses of chemicals and often left unattended, relying on infrequently maintained controls to keep them going. Some TLC (tender

Figure 6-2

loving care) once a year seems a small price to pay, given the reliable service they provide through the dead of winter.

Every boiler has its own step-by-step dismantling procedures to be followed—I suggest you consult your manufacturer's operating manual to determine the particulars for yours. Funny thing about those books, they'll tell you everything you need to know about your boiler—how to dismantle and reassemble it, what the part numbers are, even how to troubleshoot a malfunction—but it won't tell you what to look for during an internal inspection. Some assistance? Surely. At a minimum you should check for:

FIRESIDE
- flame impingement on metal surfaces
- leakage from the waterside
- integrity of superheaters/economizers
- warped, cracked or loosely seated tubes
- spalled or missing refractory
- evidence of unburned fuel

Boiler Basics

Figure 6-3a.

Figure 6-3b.

Figure 6-3. The Four Pass Construction of a Typical Cleaver-Brooks Firetube Boiler
Combustion air enters through the air inlet. The forced draft fan forces air through the rotary air damper and the diffuser into the combustion chamber. The main firetube or combustion chamber constitutes pass one. Baffling allows gases to pass to the front of the boiler only through pass two; here a baffle allows gases to pass to the rear of the boiler only through pass three. From the rear, the gases are forced through pass four to the vent. *Photo courtesy Cleaver-Brooks.*

- bulging/blistering of heated surfaces
- missing/broken baffles in gas passes
- corroded surfaces in breeching/stack
- fouled/sooted surfaces

WATERSIDE
- scale accumulation on surfaces
- evidence of oil in the water
- loose/broken/missing stays
- cracks in ligaments between tubes
- erosion/thinning of metal
- integrity/positioning of internals
- gasket seating surfaces
- blockage of inlet/discharge piping
- deformation/discoloration of metal
- evidence of pitting/general corrosion

On the other side of the coin, much can be ascertained about your boilers ability to perform by observing it in action. Even if your insurance company doesn't require the annual internal examination to be done, it's in their best interests to minimize their risk exposure; subsequently, they'll often call for a regular "operating inspection," during which their representative observes your operator putting the unit through its paces. If they don't require it—you should! *As* part of the inspection procedure, he/she may call for your operator to:

- lift the safety valves by hand
- test the low water fuel cut-off
- blow down the boiler/water column
- activate the feedwater regulator
- check the operation of the flame scanner
- run the unit on an alternate fuel
- operate the soot blowers

Sound Operating Principles

I don't know that an all-encompassing treatise of sound boiler room procedures has ever been written. For that matter, considering ever-advancing technological growth and the operating diversities of individual plants, I'm not so sure one can be; ergo, in lieu of directing you to the "perfect" reference, and given our considerable, combined field experience, I thought we might begin construction of our own. Here's my contribution...

1) Before putting a system into service, all instrumentation should be thoroughly checked out, control devices subjected to a dry run and interlocks tested for proper sequencing.

2) Once "on line," system controls and safety devices should be periodically tested, correcting any defects as they are found.

3) When purging hot gas and combustion air passages, air should be introduced at a volume and velocity sufficient to clear dead spots located along the entire length of the unit.

4) Firing rates during "start up" should be controlled to avoid excessive breeching temperatures until a flow is developed through stack mounted boiler appliances and to avoid undue thermal stress on the generating tubes and refractory.

5) Oil storage tanks should be sounded frequently and checked for water and/or sludge accumulations. Strainers should have orifices smaller than the drillings in the burner tip and should be cleaned regularly.

6) Refiring of furnaces after failed attempts should be delayed until complete purging of the entire length of the gas passage has been assured with fresh air and it is determined that no physical blockages exist along the way.

7) Access doors in the setting should be bolted or locked closed to prevent accidental opening when setting is under pressure.

8) New steam generators or units which have been out of service for an extended period of time should be subjected to a hydrostatic test of 1 1/2 times their design pressures.

9) When filling a steam boiler with water, its temperature should always be as close to the temperature of the metal of the drums and headers to avoid undue thermal stress and leaks in rolled tube joints.

10) New refractory should be allowed to dry out thoroughly and slowly to avoid damage during startup.

11) Prior to picking up the load, it is desirable to keep the water level near the lowest safe level to accommodate expansion as the firing rate increases.

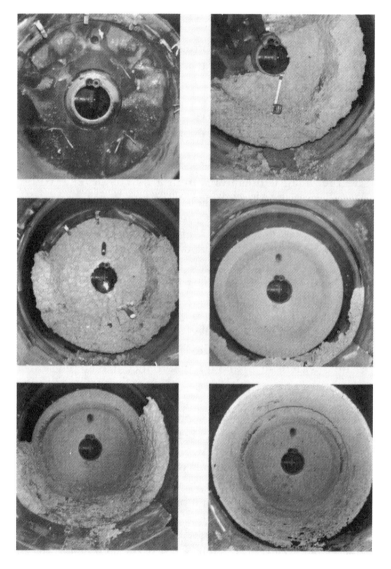

Figure 6-4. Plastic Refractory Installation on Scotch Type Boiler

12) In the case of either low or high water, the unit should be shut down immediately, the underlying cause found and corrected before returning the unit to service.

13) Water gage glass assemblies should be kept clean, well illuminated and preferably be designed such that the actual level contained in the glass is visually obvious.

14) Solid concentrations in the boiler water should be monitored daily and controlled accordingly with adequate blowdown.

15) When small leaks occur, their source should be located and repairs made as soon as the unit can be taken out of service. Serious leaks call for immediate shut down and examination by an authorized inspector before resuming operation.

16) Heat transfer surfaces should be maintained clean and in good repair at all times.

17) Soot blowers should be checked regularly for proper alignment to avoid tube cutting.

18) At the end of each heating season, boilers should be drained, cleaned, inspected and laid up in accordance with manufacturers instructions.

19) To prevent the baking on of sludge, draining of the boiler should be deferred until after it has cooled down sufficiently to allow personnel to enter and remain in the furnace.

20) Repairs should never be attempted on pressurized, energized or unisolated equipment and should only be performed by qualified individuals in accordance with A.S.M.E. (American Society of Mechanical Engineers) guidelines.

General Room Requirements

As important as proper operating procedures are, they are by no means the only concerns we have down here in the "pit." When considering the ergonomics and safeness of our operations, we often overlook the very items staring us right in the face. It's not often we give much thought to such mundane worries as exits, ways of travel and pressure vessel clearances but they are nonetheless just as, if not more important than those things we've already discussed. Here are a few for your immediate edification; you may want to consult with your local industrial board to see if they've written their own standards...

Exits

Two well ventilated means of egress leading to grade and located reasonably remote from one another should be provided for —

- boiler rooms containing high pressure boilers and blow offs

- high pressure steam line tunnels

- boiler rooms of over 600 square feet of area containing low pressure boilers

- other places where there is a danger that employees may be trapped in confined spaces in case of explosions or steam line bursts

Ladders and Walkways

All platforms and walkways should —

- be of bolted or welded metal construction

- be provided between or over top of boilers which are over 8 feet high

- be constructed of safety treads or standard grating with a minimum 30-inch width

- be equipped with 42-inch-high handrails with an intermediate rail and 6-inch toeboard

Ladders which serve these platforms and walkways should —

- be of metal construction; not less than 18 inches wide

- have rungs extended through the side members and permanently secured

- have a distance of not less than 3 inches from the front of the rungs to the nearest permanent object on the climbing side

- have a distance of not less than 6-1/2 inches from the back of rungs to the nearest permanent object

- have a clear width of 15 inches from the center line across the front of the ladder to either side

Clearance

The minimum clearance around and between equipment and ways of travel should be —

- 30 inches between and around boilers

- at least 6 feet from the floor to any overhead obstruction

- 18 inches between unfired pressure vessels except in front of manhole openings (30 inches)

- 12 inches between the floor and underside of an unfired pressure vessel

- 6 inches above the highest point of any valve or fitting

- 7 feet between boiler platforms and ceilings (minimum of 3-1/2 feet from top of boiler to ceiling)

Increasing Operating Efficiency

Too often, engineers think of operating efficiency only in terms of fuel consumption at the burner and/or percent increase in make-up feedwater. The reality is, there are a multitude of areas that are often overlooked which, if properly addressed, would greatly improve the overall equipment efficiencies of your plant while significantly reducing its operating costs. I'll list a few for you to maul over...

Combustion Efficiency
- reduce excess air to the furnace
- adjust burners for optimum consumption
- convert to more efficient fuels
- switch from steam to air atomization
- check boiler air casing for leaks

Waterside Efficiency
- decrease continuous and/or bottom blowdown
- install blowdown heat recovery system
- repair leaks in condensate return lines
- preheat make-up feedwater

Heat Transfer Efficiency
- clean boiler heating surfaces
- discontinue on-off operation
- perform daily feedwater analysis
- maintain baffles in good repair

Steam System Efficiency
- shut off steam traces during mild weather
- repair/install insulation where needed
- repair and/or replace faulty steam traps
- restuff valve and pump packing glands
- repair leaks in lines and reducing stations

Chapter 7

Chiller Essentials

The term Boiler Room is often used interchangeably with Power Plant. Though the steam generators residing there often serve as the base of operations, as previously intimated, boilers are only a fraction of the overall power plant equation (the heating side). Equal weight must also be given to the Chiller Room (the cooling side). And the two of them together only comprise 50% of the total, when you take into account the auxiliaries in the Mechanical Rooms and the distribution panels/transformers in the Electrical Room. But more on those hot topics later. For now we're strictly playing it cool...

System Component Variations

Refrigeration systems, as you are most probably aware, can be classified as either absorption or compression by type. Although the former certainly has its place in the physical plant, it's the latter, more familiar compression type that we'll be addressing in this writing.

Mechanical compression systems can be categorized as either centrifugal or reciprocating, depending on their mode of compressing the system's refrigerant. Positive displacement reciprocators provide compression through the action of a piston squeezing refrigerant vapor in a cylinder, whereas variable displacement centrifugals have rotating impellers than impart centrifugal force upon the refrigerant gas, resulting in the head necessary for the refrigeration cycle. These systems can be further divided into open and closed (hermetic) groupings, depending on the housing of the drivers of their compressors and classified as either direct expansion, where liquid refrigerant is delivered directly to the cooling coil or as indirect, where the refrigerant is used to extract heat from a secondary coolant, such as brine, which is circulated through the cooling coil.

However different their designs, all systems conform to the same operating principles in reference to the basic refrigeration cycle, and after the issue of compressor type has been decided, system components become interchangeable. As can be seen in Figure 7-1, the four main components involved in the refrigeration cycle are the compressor, condenser, evaporator and refrigerant expansion valve. As we've already covered the basic differences between compressor types, that leaves us with these components to differentiate:

Condensers (3 types)
- Air Cooled
—use propeller or centrifugal fans to move air across fin tube heat transfer surfaces

- Water Cooled
—water is passed through shell and tube heat exchangers to absorb heat from the refrigerant vapor in contact with the opposite surfaces. The cooling water is then either disposed of or recycled through a cooling tower.

- Evaporative
—refrigerant gas is passed through coiled tubes, the surfaces of which are sprayed with water as air is passed over them in an upward direction to induce heat extraction by evaporation

Evaporators
- Fin Tube
—air is blown or drawn past a finned coil through which refrigerant is circulated.

- Shell And Tube
—liquid refrigerant enters through one end of the heat exchanger vaporizing as it extracts heat from the fluid in contact with the opposite side heat transfer surfaces.

Chiller Essentials

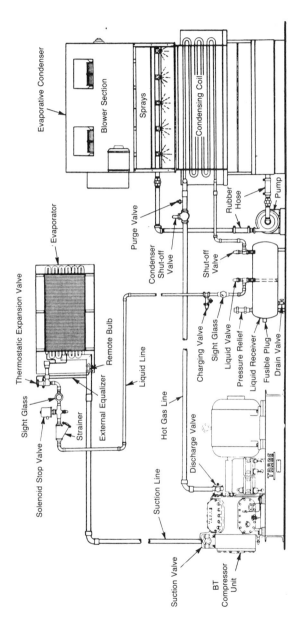

Figure 7-1

Refrigerant Metering Devices

- Automatic Expansion Valve
  —modulates refrigerant feed to maintain a constant preset pressure in the evaporator

- Thermostatic Expansion Valve
  —measures the condition of the refrigerant vapor leaving the evaporator (the degree of superheat) modulates the flow of refrigerant liquid to the evaporator

- Capillary Tubes
  —lengths of small diameter tubing separating the high and low pressure sides of a system, having a fixed orifice through which the refrigerant is fed to the evaporator.

Mechanical Refrigeration Standards

When the time comes for you to get serious about equipment room design, say if you're installing a new system or performing major surgery on your old one, you should refer to ASHRAE (The American Society of Heating, Refrigeration and Air Conditioning Engineers) Standard 15-1992 for assistance. This standard, "Safety Code for Mechanical Refrigeration," is intended to promote the safe design, construction, installation and operation of refrigerating systems and establish reasonable safeguards of life, limb, health and property. It is a comprehensive informative document, the contents of which, from a Facility Management perspective, might be described as: everything you wanted to know about mechanical refrigeration systems but were afraid to ask! It's a veritable "soup to nuts" coverage of mechanical refrigeration in every conceivable regard. In it you'll find references to pressure relief piping, use of open flame devices, long-term monitoring, clearances, access, egress, alarms, ventilation... Did I say new system installation? Hell, you can't go wrong utilizing this standard for your day-to-day operations. Can you think of a better document from which to draw information in formulating your operating policies and procedures?

Since its inception, ASHRAE has been an indispensable informational source for all proponents of the heating, refrigeration and air-

Chiller Essentials

Figure 7-2

conditioning trades. Enough can't be said of the contributions they continually make to us. If you're not already familiar with the organization, what they're about and have to offer, it would behoove you to become so.

Installation Specs

You don't have a copy of Standard 15 in your hip pocket at the moment? Obviously its possession isn't prerequisite to writing a refrigeration or air-conditioning installation bid spec, but I do highly recommend it as a guide. In lieu of same, here's a skeletal template to get you headed in the "write" direction. You can expand on it as need be.

I. Scope of the work
   a) detail the system and components to be installed
   b) establish a time frame for partial and total completion
   c) stipulate compliance with schedules, plans, drawings and applicable codes
   d) determine acceptable workmanship and materials quality
   e) delineate "who" is responsible for "what" and arrange estimated delivery times/dates

II. Insurance and Permits
   a) determine responsibility for acquisition, payment, types and amounts to assure total compliance
   b) spell out all coverages

III. Materials Specification
   a) type of piping/fittings including wall thickness
   b) types of refrigerants and reclaimability
   c) brand names of accessories to be utilized

IV. Description of Work
   a) how everything should be installed
   b) where everything will go
   c) what special considerations will be required
   d) in what order work should be accomplished
   e) special methods and techniques to be utilized

V. Approvals
   a) location and type of equipment
   b) accessibility for repairs
   c) use of subcontractors
   d) substitution of materials

VI. System Commissioning
   a) evacuation and charging parameters
   b) testing of all components under load conditions
   c) repair of leaks and replacement of malfunctioning devices
   d) complete operational check out and report
   e) tagging and labeling of valves and piping by direction of flow and area served

VII. Documentation
   a) manufacturers instruction manuals
   b) operating logs and data sheets
   c) certificates of operation
   d) as-built drawings/close out documents
   e) warranties and guarantees

Assuring Operating Integrity

The most complicated component part of a mechanical refrigeration system, and the one most likely to fail, is the compressor. But it isn't its complexity that causes its early demise. Seldom do compressors experience the rare luxury of wearing out as the cause of normal operation. History has shown that most failures occur, not as the result of internal malfunctions, but rather as a consequence of problems suffered elsewhere in the system which ultimately affect compressor performance.

Studies have also evidenced that in almost every instance, there were clear indications that problems existed in the systems prior to their shutting down. No better argument can be made for the need of a preventive maintenance program, and these are the items you should be keeping an eye out for:

- log entries made by operating personnel reporting abnormal or erratic operation, pressures and temperatures

- improper oil and refrigerant levels, condition and coloration

- wearing, cutting or broken parts resulting from excessive vibration

- noisy malfunctioning and/or dirty controls

- excessive sweating or frosting of component parts and lines

- insufficient insulation, ventilation and lubrication

- inappropriate environmental conditions and filtration

Psychrometric Observations

Psychrometry is defined as the science dealing with the physical laws governing the behavior of air-water mixtures. This is different from psychiatry, which is the science utilized to unscramble our brains after attempting to decipher our psychrometric charts.

Due to its ability to absorb moisture, air has five variable characteristics which affect its "condition" at any given time. These properties can be recorded on a special graph called a psychrometric chart, which enables the analysis and manipulation of these properties in "conditioning" the air. The five properties are:

1) Dry Bulb Temperature
   - read from an ordinary thermometer which has a dry bulb

2) Wet Bulb Temperature
   - read from a thermometer whose bulb is covered by a wet wick

3) Dewpoint Temperature
   - the temperature at which moisture in the air condenses onto objects it surrounds

4) Relative Humidity
   - the ratio of the amount of moisture in the air to the amount it is capable of absorbing

5) Humidity Ratio
   - the actual weight of water in a mixture of water vapor and air expressed in grains of water per pound of dry air

The beauty of the chart is in its ability to provide the three remaining values when any two of the five are known. More sophisticated charts such as that represented in the Figure 7-3 skeleton chart (on page 110), provide additional information for upper scale engineering work. Can you pick out the five lines that depict the five basic properties we've been discussing?

Cold Weather Operation

Being the frugal operator that you are, I know you're going to want to take advantage of some of that so-called "free cooling" available to you from your cooling tower during the winter months. Your quest to conserve energy (bucks) is applaudable and you should be commended for expanding your engineering base, but make sure you do your homework before attempting to "modify" your system.

Essentially, there are two ways of acquiring thermocycle cooling within a chilled water system, both of which require reworking of your systems configuration. One involves bypassing of the refrigerant around the compressor and circulating unusually cold water through the condenser causing the refrigerant to condense at a lower pressure/temperature than normal. The subcooled refrigerant is passed through a PRV station and then an expansion valve producing a refrigeration effect. The refrigerant vapor provides the horsepower for recirculation.

The second involves cross connection of the cooling tower and chilled water circuits and rendering the entire refrigeration side of the system inoperable. Water from the cooling tower is then circulated through the chilled water loop to the same components normally served by the intact system. Both of these operating modes require physical reworking of system hardware, structural reinforcement of cooling tower elements and critical control of operating temperatures. Though it has been proven that these modes of operation can provide cost and energy savings, the extent and pay back largely depend on the weather, frequency and quality of the increased maintenance required, as well as the specifics of the users particular system design. That being the case, I'm sure you'll agree that the prospect should be thoroughly researched before attempting an in-house "fix."

Whether or not you are availing yourself of the foregoing operating advantage, it may be necessary for you to provide a cooling capability to all or a portion of your buildings via traditional operation of your

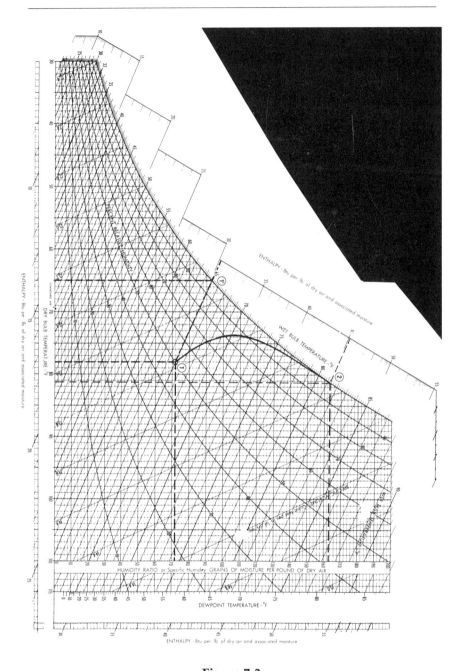

Figure 7-3

chillers during periods of cold. When such is the case, consideration should be given to the following:

- antifreeze protection

- reduce fan speed to increase water outlet temperature

- restrict air intake openings

- frequently assure operation of heating elements

- install a hot water bypass to the cold water basin

- de-ice surfaces before putting into operation

- drain exposed piping and water basins prior to prolonged shut downs

- decrease the number of operating cells to increase the heat load over those remaining

- use heat tapes on exposed piping

- thermostatic control of fan operation

- operate tower at maximum heat load

- frequent visual inspections and stepped up maintenance intervals

- maintain temperature of circulating water as high as operationally practical

- remove ice by operating fans in reverse direction for short periods of time

- correct fan imbalances caused by ice accumulations using steam or hot water

- P.S. Don't forget the ladders, railings and walkways... (operator safety is as important as operating safety)

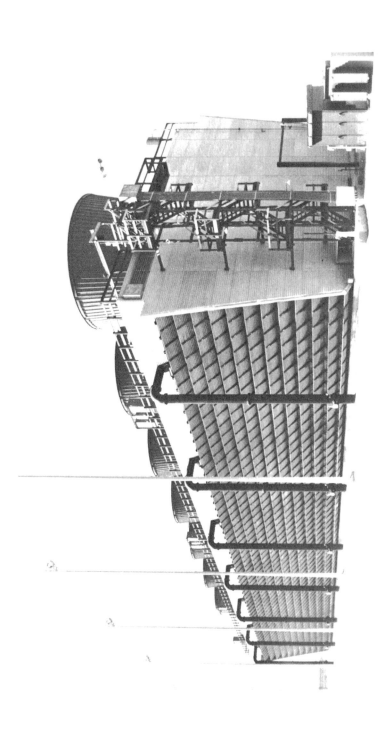

Figure 7-4

CHAPTER 8

AUXILIARY/SUPPORT EQUIPMENT

The project completion date is drawing near and the contractors are chomping at their bits in anticipation of turning over the responsibility (i.e., liability) for the building's equipment operation into your capable hands. You're still not an expert on every device? Don't worry... be, uh, become better informed! Considering the functional diversity, types and sizes of the equipment found in today's power plants, no one knows it all or can do it all. Any even if you're extremely sharp in some regards, it doesn't hurt to be well-rounded in most. To that end, here are some ditties I've assembled for your edification.

Compressor Accessories

When it comes time to install compressors in our plants, too often we think of them only in terms of a desired end result—that of supplying vapor, under pressure, to a piping distribution network. What we have a tendency to overlook is that compressing units are more than just individual pieces of equipment performing a particular function, rather they are self contained systems of which the compressor is the heart but not the only part. The compressors "accessories" are components, the function of each adding to the safety and reliability of the overall system. Don't forget these when specifying your system components:

- compressor intake filters
  (dry type? oil type? viscous - impingement)

- dryers
  (chemical, mechanical, refrigeration type)

- traps and separator
  (moisture removal, oil separation, etc.)

- intercoolers and aftercoolers
  (for removing heat of compression, reducing temperature and moisture content)

- lubricators and cooling water systems (types and proper additives)

- operating controls and safety devices
  (actuators and trip outs with appropriate manual and automatic resets)

- pressure/temperature gauges (operator monitoring and equipment operating history)

Cooling Tower Structures

Look—up in the sky—it's a bird!—it's a plane!—it's... I'll be damned —it's the sun! Now that may not be an apocalyptic revelation to you fellows in the more southerly climes, but for us it means we can go up on the roof and check out the structural integrity of our cooling towers without freezing our fannies off in the process. And, even though there's still some frost on the pumpkins, we know all too well, from past experience, that the cooling season will be on us before the water hits the ground when it melts. With that in mind, why don't we lock out the disconnects, climb up onto, around and through the tower and...

Check the exterior for:

- missing or broken access doors/panels

- the number of louvers missing or damaged or mispositioned

- decay and dry rot of wooden components

- squareness of corners and evidence of leakage

- deteriorated drain boards, splash guards, water troughs and distribution basins

- structural stability of ladders, walkways and fan decks

- evidence of metal fatigue and hardware detachment

- corrosion of pipes, valves and flanges

Check the interior for:
- evidence of corrosion and acid attack

- splitting, warpage, alignment and attachment of interior components

- missing or plugged nozzles and distribution holes/devices

- deterioration of drift eliminators

- evidence of algae and debris accumulations

- soundness of braces and supports

- general condition and cleanliness of screens and surfaces

- corrosion of hangers, pipes and connections

- missing or deteriorated fasteners and gaskets; sagging or broken parts

- operating integrity of fire warning and suppression system components

Electric Motor Selection

If you're rocking in the cradle somewhere between a thoroughly dilapidated plant and a totally new one, selection of the proper electric motor for a given application is fairly easy. If the old motor you're

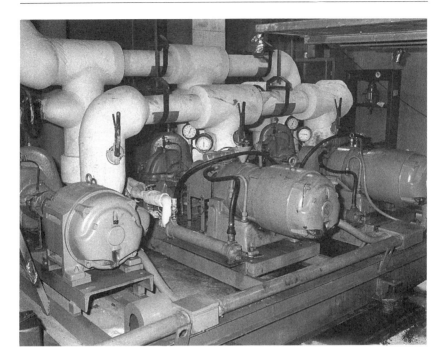

Figure 8-1

replacing did the job, all you need do is remove the existing one, substitute an exact duplicate and voila... you're back in business! On the other hand, if the nameplate is missing or worse yet, if the motor itself is missing, you've got a problem. A rose, as they say, by any other name may still be a rose, but an electric motor can be classified in many different ways. It isn't enough to know you need a 25-horsepower motor.

Depending on where it's run, for how long, how often, under what conditions, etc. it will need to meet certain operating criteria. At a minimum, decisions will be needed as to type, size, speed, duty, kinds of bearings, type of enclosure and the mounting base to be utilized.

As you can see, before a proper motor can be selected, many questions need answering, not the least of which is... load rating? Load ratings deal with the starting conditions to which a motor is subjected. Driven units of various types, once in operation, may require the same horsepower motor to run them, but getting them started is sometimes another matter. Often machines require special torque capabilities to bring them from a full stop up to normal running speed. In the plant, three common classes of starting conditions you encounter are:

Unloaded Starts
(load completely blocks motion after the machine stops)

Partly Unloaded Starts
(reciprocating gas and air compressors)

Fully Loaded Starts
(machines may be damaged if started unloaded)

As previously intimated, there is much for you to know about the motors you select for a given application. For example, in-depth studies will determine if your motors design should be general, specific or special purpose, into what insulation class they should fall and/or the correct plane in which they should operate based on bearing design. But typically, most power plant managers can get the job done if they know:

- The proper type based on the motor's starting mechanism
  (single or multi-phase)

- The proper size based on how much power is required to drive the load
  (undersized motors overheat and burn up - oversized motors draw excess current increasing operating costs)

- The proper speed based on the desired end result
  (the higher the rpm, the quicker the movement, the more air that's blown, the more water that's pumped, etc.)

- The proper duty based on loading duration and frequency of operation
  (continuous or intermittent)

- The proper bearings required based on mounting position, method and frequency of lubrication
  (sleeve, ball or roller types—sealed, hand packed or fitted)

- The proper enclosure based on environmental and atmospheric conditions
  (open, totally enclosed and explosion proof)

- The proper <u>mounting</u> based on fit-up, alignment and vibration parameters
  (fixed or adjustable, rigid and cushioned bases)

Generator Maintenance

Except for those of us who are 'into' the co-generation thing, or are otherwise, required by an accrediting body or regulatory statute to test our back up generators on a schedule, we tend to forget just how important these machines are to our operations. During interruptions to the normal power supply, they literally save lives, eliminate crime, alleviate fear, maintain productivity levels and enable the continuation of the normal order of things. I'm impressed! So why don't we take better care of these units on the whole? Beats me. You can imagine the flip side of the scenario. By all means, construct yourself a thorough PM procedure to care for them but by no means have it provide less attention than:

<u>Daily</u> checking of the rooms in which they are housed to assure proper temperature, louver function and freedom from clutter and debris. The generator breaker should be in the "on" position and the panel selector switch in the "auto" position.

<u>Weekly</u> inspection of the batteries for charge, oil heater for function and belts for tension

<u>Monthly</u> determination of electrolyte, oil and coolant levels, position of air cleaner indicators and operation of the generator under full load conditions

<u>Semi-Annually</u> change oil filters and governor oil, lubricate external moving parts and inspect all cables and linkages

<u>Annually</u> remove and clean battery connections, check all adjustments per manufacturers recommendations, replace belts, and filters, drain and clean sediment from tanks and pans

<u>Bi-Annually</u>   Flush cooling system and replace all hoses, tune and adjust engine and check exhaust system for leaks

Figure 8-2
B&G large vertical split-case centrifugal pumps installed in new commercial office building. Compact vertical split-case pumps save valuable floor space and easily fit into crowded equipment rooms.

Pump Repairs

Second to electric motors, pumps make up the bulk of the support equipment inventory. True, they supply fluids through the miles of piping hidden in your walls, floors and ceilings but that's not their only function. They recirculate hot water in loops to maintain temperature, boost line pressures on upper levels, transfer chemicals from mixing tanks to make-up tanks, discharge sewage from sumps, maintain pressurization of wet sprinkler systems... the list continues *ad nauseum*. But whereas the pump places second with regard to total numbers, it has proven itself to be just as reliable a machine as the electric motor, if not more so, considering the abuse it is constantly subjected to. But unlike a Timex, if you give it too much of a licking, it will most assuredly quit ticking—at the worst possible time. The nature of the vexations? Sure.

The most common of all pump repairs is the replacement of worn or broken parts. A supply of manufacturer recommended spares should be kept in stock, as should frequently replaced items particular to your operation, as well as consumables utilized in your preventive maintenance program. Some of the more common repairs include:

Impellers
Wear in small units is best corrected by replacement. Cracks in larger units may be rebuilt by brazing, soldering or welding

Bearings
Worn or damaged sleeve bearings may be rebabbitted. Defective ball and roller bearings should be replaced.

Casino
Cracks in casings can be repaired either mechanically or by welding.

Wearing Rings
Casing and impeller rings are designed to wear and should be replaced as indicated in the pumps operating manual.

Mechanical Seals
All seals have two flat seating surfaces mounted perpendicular to the rotating axis. If these surfaces are not absolutely flat, the misalignment will shorten the effective life of the seal resulting in premature failure.

Packing
Don't continue to add new packing to old/worn packing. Completely clean out all old packing and replace with new, fresh rings.

Shafts
Scored shafts can be metal sprayed and machined, but replacement of the shaft is always preferable.

Flexible Couplings
Flexible couplings are designed to compensate for shaft length changes caused by variations in temperature and to prevent axial thrusts be-

tween the driver and pump shafts. They will not compensate for misalignment.

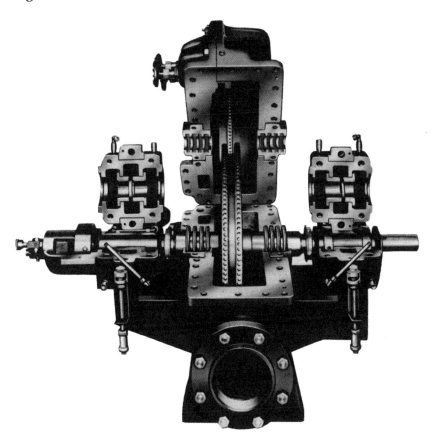

Figure 8-3

Turbine Inspections

Whether they're generally misunderstood, considered the sovereign province of the operating engineer or just so dependable that they're easily forgotten, turbines don't make it into the pages of technical references nearly as often as they should. But as reliable as they are, they pose a high economic risk should they fail. Subsequently, I've decided to pay some attention to this critical component here. I'm sure you'll take full advantage of whatever information is available to you by way of the manufacturer, but if you're in a pinch to come up with your own set of guidelines, try these checklists on for size.

Operating Inspections

To determine the safety and operating integrity of a unit in operation, monitor its speed, loading, vibration and check out these items and/or conditions:

Bearings
- thermometer operation
- abnormal temperatures
- lube oil supply
- oil levels and pressure

Condenser
- loss of normal efficiency
- dependability of water supply
- overpressure protection
- evidence for leaks
- deterioration of piping

Controls
- governor linkage and response
- main admission valve and gear
- cams and rollers
- valve stems and packing
- rack and pinion gears
- extraction valves and gear
- lifting beams and bars

Drains
- exhaust line
- separator
- throttle
- main steam line
- bleeder line casing proper location evidence of leakage general condition

Foundation/Casing
- cracking and leakage
- settling, movement or looseness
- freedom of expansion
- lubrication/movement of pedestal
- deteriorated grouting

Lube Oil System
- oil pump/cooler operation
- pump/governor/bearing pressures
- bearing/cooler temperatures
- lube oil sump drain
- oil condition/filtration

Oil Leaks
- governor system
- lube oil piping/fittings
- pump glands/bearing seals
- valve piston rod glands

Piping
- cracks and leakage
- alignment/expansion
- supports and hangers
- overpressure protection vibration

Safety Devices
- overspeed governor
- non-return valves
- throttle trip valve
- overpressure relief valves
- temperature/pressure cut-outs alarms

Steam Leaks
- casing joints
- piping/connections
- valve packing
- gland seals

Turbine Auxiliaries
- air ejector
- vacuum pumps

Dismantled Inspections

Just as important as it is to witness the credibility of a turbine's operation, so too is the observation of its internal workings during periods of down time. In conformance with the frequency suggested by the manufacturer, these units should be dismantled, inspected and overhauled. This activity should include:

- inspection of the steam chest and nozzle blocks

- inspection of the top/bottom halves of the turbine casing

- dismantle and examination of the governor/overspeed governor assemblies

- removal/cleaning of the rotors and examination of rotor component train

- check and clean components of the main and auxiliary lube oil systems

- visual inspection of throttle, emergency trip and non-return valves

- determination of appropriate clearances between stationary and moving parts

- removal/testing of safety valves/devices

- examination of the journal and thrust bearings

- non-destructive examination of critical parts for discontinuities

- visual inspection of steam glands, seals, packing, strainers and drains

- removal and checking of all control valve components

- examination of pedestal, foundation and coupling

- replacement of oil seals and worn or defective parts

CHAPTER 9

ELECTROMECHANICAL ROOMS

Until now, we've explored this place in a forthright manner, covering things up front and out in the open. But there's a seamier side to its existence. Like confessional boxes, its mechanical and electrical rooms are receptacles for the multitude of sins committed by the technical disciples we send forth into them. Tucked away in the dark recesses of our buildings, the less glamorous of our electromechanical devices are often neglected, while homage is paid to the systems with the bells and whistles in the well lit rooms.

Adhering to the tenet that "God helps those who help themselves," might we be able to redeem ourselves by caring for our less visible items and areas say... more religiously?

Air Handling Units

There's more to the ducting and grillwork overhead than knowing whether it sucks or blows. Is it connected to an exhaust or supply fan? With the direction of flow established, is the volume adequate? Characteristics correct? Pressure proper? Is the delivery unit intact and monitored? Support equipment maintained? Are conditions recorded? Is equipment cycled to conserve energy? Are repairs completed in a timely fashion? Did I air out my point well enough? At the least, here's what you should know about your air handlers:

- system types (low pressure - constant volume? high pressure - dual duct? variable air volume?)

- percentage of outside air (return air damper settings... utilization of reheat)

- types of fans and performance curves (ability to change CFM's, fan testing, troubleshooting, procedures...)

- drive sheaves installed (fixed/adjustable? belt tension, sizes and alignment)

- Ak area of grilles and diffusers (unobstructed air flow area corrected for physical restrictions and measuring instruments)

- associated instrumentation (anemometers, thermometers, psychrometers, humidistats, magnehelic (draft) gauges, velometers, Pitot tubes...)

- air balancing requirements of supplied areas (number of air changes, positive/negative pressure requirements...)

- preventive maintenance frequencies and type
  (care and cleaning of system components and drains)

- documentation of operations (air balancing attempts, equipment change-outs, operating temperatures/pressures...)

Electric Power Systems

We building engineers generally agree that of all our systems, those supplying electrical energy are the most critical and at the same time, the most dependable. It may come as a shock to you (sorry, I just had to do that), but these systems are also the most neglected. Being the reliable partners they are we, often take them totally for granted, as we sometimes do a close friend. But unlike a faithful cocker, these puppies don't whimper, they bite! How do you muzzle them? Through proper understanding and maintenance of your systems. Here's what you need to know and do:

- trace your system from the utility source to the end of your branch lines to learn:

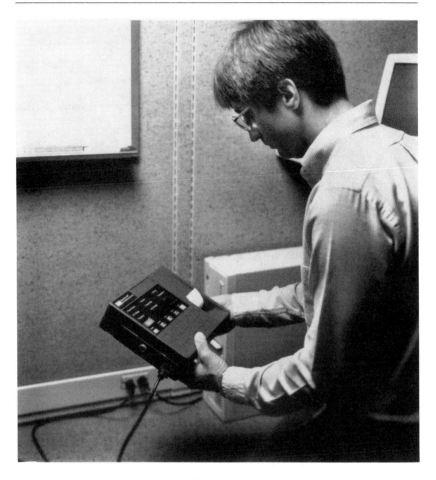

Figure 9-1
The PowerVisa is the only power monitor that detects, documents, and explains power problems, especially for the person who knows little about them.

—what's the incoming voltage?
—how many substations feed your installation?
—who owns the transformers?
—how reliable is the power? the service?
—what are the primary values? secondary?

- check out the safety and reliability of your systems for proper...
    —grounding and lightning protection?
    —installation of ground fault interrupters

—single phase protection? protective relaying?
—uninterruptable power supplies?
—back up emergency power?
—fire suppression systems?

- determine whether regular inspections and frequent maintenance of your systems is scheduled for...
—performance of thermographic studies?
—meggering of insulation?
—tightening of loose connections?
—exercising of moveable parts?
—evidence of worn/burned components?
—changing of oils and filters?

- be aware of what's going on inside your systems by establishing the integrity of...
—panelboard indicator lights?
—circuit service identifications?
—power measurement instrumentation?
—phase rotation? power factors?
—system imbalance correction practices?

Uninterruptible Power Supplies

While we're on the subject of electrical systems reliability, we shouldn't overlook uninterruptible power supplies. Utilizing storage batteries as a power source, UPS systems maintain specific power requirements of selected equipment when the normal (utility) supply is halted or falls below minimal levels', thus they can be used to assure the continuity of electrical power to devices or condition it to meet predetermined values.

Such systems are generally classed as either ROTARY or STATIC by type, depending on their construction. Rotary systems utilize a motor-generator and power inverter (converts DC to AC) with battery and charger. Normal power drives the MG set which conditions the power it in turn supplies. When normal power goes out of tolerance, that duty is transferred to the battery powered inverter. The inverters may be either on-line or off-line type. Static systems depend on rectifiers (converts AC to DC) in place of the MG sets used in conjunction with inverters to reconvert the DC power back to AC. Both manual and emergency power flow through the inverter. Which of the two types to select is a function of the application for which it is installed. As a rule of thumb, static systems are less costly, more efficient, smaller and lighter than comparable rotary systems.

Systems should be selected based on the type and frequency of anticipated power outages, the effect and cost of equipment downtime, the sensitivity of your electrical devices, the availability of space for housing them, but most importantly their maintainability, which should include:

- monthly inspection of the batteries and specific gravity/voltage readings of their cells

- daily recording and analysis of system voltages/currents

- annual infrared scans, tightening of connections, and exercising of circuit breakers

- regular calibration of instrument monitors and testing of system alarms

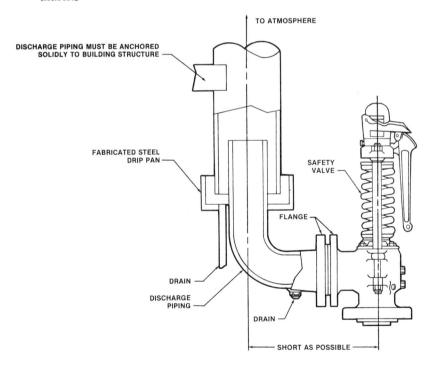

Figure 9-2

Pressure Relief Valves

The last bastion of defense against overpressure conditions in your containment vessels will most likely be a rupture disk, fusible plug or similar device, which upon activating completely dumps its contents. But long before such a final solution becomes necessary, your operating/safety controls should have alleviated the problem and automatically cycled the unit or shut it down. In between those two extremes comes the activation of pressure relief valves, which reduce accumulated pressures down to pre-determined levels whereupon the valves reseat themselves until their lifting pressures are once again attained.

The four types commonly found in our power plants are the SAFETY VALVE/RELIEF VALVE, SAFETY RELIEF VALVE, PRESSURE/TEMPERATURE RELIEF VALVE. These valves can be described as follows:

Safety Valve

An automatic pressure relieving device actuated by the static pressure upstream of the valve, characterized by a "pop" action (sound) and full, rapid opening, allowing pressure to discharge through the full cross sectional area of the outlet. They are used in steam, gas or vapor service on boilers, air tanks, piping, etc.

Relief Valve

An automatic pressure relieving device actuated by the static pressure upstream of the valve, characterized by slow, ever increasing opening of the valve in proportion to the increase in pressure over the opening pressure. They are used primarily for liquid service on hydro-pneumatic tanks, water lines, pumps, etc.

Safety Relief Valve

An automatic pressure-actuated relieving device suitable for use as either a safety or relief valve depending on application. For liquid service, the set pressure is considered to be the inlet pressure at which the valve starts to discharge. For vapor service, the set pressure is considered to be the inlet pressure at which the valve "pops."

Pressure/Temperature Relief Valve

An automatic pressure relieving device actuated by the pressure and/or temperature of a vessel's contents. Used primarily on hot water heaters, when pressures reach the valve setting or temperatures approach the boiling point, the valve will open slightly to pass some liquid or a fusible element will melt to relieve overpressure.

These valves should be tested frequently in accordance with manufacturer's recommendations. To avoid premature failure due to

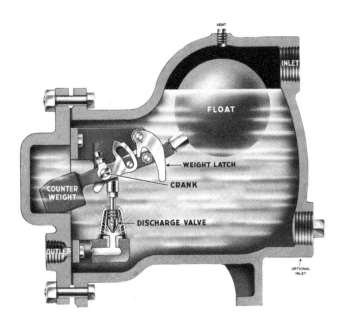

Figure 9-3

How it works. The Model JR Trap operates on an entirely different principle than ordinary "float" traps. In operation, the condensate raises the float to its highest point of travel, releases the weight latch, allowing the counterweight to fall which opens the discharge valve fully and instantaneously through a crank mechanism. A link latch holds the valve wide open until the float descends to its low point where the weight latch engages the counterweight. The link latch then disengages, closing the discharge valve instantaneously. The rapid opening and closing of the valve prevents wire drawing of the valve and seat. A liquid seal is maintained over the discharge valve, preventing the loss of steam, air or gas.

build-up of foreign deposits on valve disks and seating surfaces resulting in weeping and/or sticking problems. Careful attention should be paid as to types numbers, pressure settings, calibration, relieving capacity and points of discharge.

Steam Trapping

In the decades I've spent wandering through scores of power plants, I have yet to find one where the steam traps were properly placed. Too often they're installed in out-of-the-way/hard-to-reach, nooks and crannies, neglected and forgotten until they hammer their way into your heart, mind and wallet. If you have anything to do with the laying out of your condensate return system, make certain to locate your traps such that they are easily accessible for inspection/repair, and they are fitted with the proper shut-off valves, unions and strainers.

The three basic traps commonly found in most plants are the MECHANICAL, THERMOSTATIC and IMPULSE types. These traps can be described as follows:

MECHANICAL TRAP

A device using floats or buckets to discharge accumulated condensate (continuously or intermittently by design) from steam lines. This type of trap should not be used where a possibility of freezing exists.

THERMOSTATIC TRAP

A device using a liquid filled bellows that expands and contracts when exposed to changes in temperature, that's connected to a valve plug, which facilitates the discharge of condensate from steam lines. The bellows enables the trap to anticipate opening and closing requirements virtually eliminating leak through. Thermostatic traps possess superior condensate and air handling characteristics and can be used out of doors.

IMPULSE TRAP

A device which operates on the principle of pressure differential across a valve disc to discharge condensate from steam lines. These traps continuously leak through steam to keep air from creating problems and as such have fallen into disfavor with plant operators and are seldom used.

Traps can fail in either the open or closed position. Those failed in the closed position are the ones that cause all the commotion. They will be cool to the touch and usually cause waterhammer in the lines or equipment emptied by the trap. Those failed in the open position aren't as easy to find due to the lack of temperature difference between inlet and outlet as the result of the steam blowing by. Some troubleshooting tips? Sure...

STEAM BLOW BY
- valve failed to seat due to worn valve parts or scale lodged in orifice
- Trap lost prime due to drop in steam pressure

COLD TRAP
- No water or steam coming to trap
- Worn or defective mechanism
- Broken valve in line to trap
- Trap filled with dirt

HOT TRAP
- No water coming to trap
- Vacuum in coils may be preventing drainage
- Mechanism hung up

CONTINUOUS FLOW
- Trap too small
- Boiler priming

Equipment/Room Checks

However you choose to set yours up, it's imperative that you establish routines for your operators to check all this stuff out. Not just the equipment but the rooms themselves. It is advisable to check the power plant out "totally," at least once per shift. Our plan here is to schedule two engineers per shift; one will be responsible for operations in the main boiler and chiller rooms and the other for the mechanicals throughout the remainder of the plant, as follows:

STEAM TRAP SURVEY — Figure 9-4

Barnes & Jones, Inc.
34 Crafts Street
P.O. Box 155
Newtonville, MA 02160
TEL: (617) 332-7100
FAX: (617) 965-3482

329 Geneva Ave., No. #209
St. Paul, MN 55128
TEL: (612) 731-9041
FAX: (612) 731-0536

FOR BUILDING: _____
ADDRESS: _____

SEE BELOW FOR EXPLANATION OF SURVEY HEADINGS. PLEASE BE AS SPECIFIC AS POSSIBLE.

1	2	3	4	5	6	7	8	9	10
TRAP LOCATION IN BUILDING USE EITHER ACTUAL LOCATION OR TRAP TAG # (IF ALL TRAPS IN BUILDING **HAVE BEEN** SEQUENTIALLY NUMBERED FOR IDENTIFICATION)	MFG. NAME	PIPE SIZE	MODEL #	SYS. OP. PRES.	TRAP PRES. RATE	TYPE OF UNIT TRAPPED	OTHER INFORMATION	COMPLETE TRAP REPLACEMENT	REPAIR EXISTING TRAP

1. – Where trap is physically located in building.
2. – Name of manufacturer – usually located on top or side of trap.
3. – Size of pipe entering steam trap.
4. – Model number – usually located on top or side of trap.
5. – Steam pressure of unit being trapped.
6. – Manufacturers rated trap pressure – on top or side of trap.
7. – Heating unit being trapped – e.g., radiator, fan coil unit, etc.
8. – Miscellaneous information about trap performance: History of trap problems? Date last repaired? Complete or parts?
9. & 10. – Maintain service record on complete trap replacement or repair parts installation.

Electromechanical Rooms

WATCH ENGINEER (Boiler/Chiller rooms) Prior to taking over the watch, the operator coming on duty will review the events occurring on the prior shift with the watch engineer to assure himself of the status of all systems, equipment and incomplete work.

Upon taking over the shift, he will tour the boiler and chiller rooms, observing the equipment on line to determine its operating integrity.

On a continuous basis, he will:

- check the water level in the boilers

- listen for unusual noises generated by the equipment

- monitor the equipment in operation for vibrations, temperature and pressure variations, leaks and alarming

- lubricate and adjust equipment/controls as needed

Each hour during the course of the shift the watch engineer will check:

- the operation of the feedwater, sump/ejector, domestic water, chilled water, and circulating pumps

- the liquid levels of all tanks and operations of their indicators and components

- for proper positioning of all valves and switches

- the operation of all auxiliary equipment including chemical transfer units, medical gas compressors, vacuum pumps, etc.

Once each shift, the watch engineer will:

- check the hardness of the water at the softener station and add salt as needed

- blow down the boilers water columns and bottom blow down as indicated by analysis of the feedwater

- lift (test) all safety valves by hand (other than on the high pressure header and boilers)

- pull samples of boiler, cooling tower and condensate waters, analyze and record findings

- add chemicals as needed to chemical feed pumps as indicated by water tests

- "bump" oil transfer pumps to assure operating capability

- "rotate" all motor shafts by hand on unenergized equipment

- turn all written documentation into the Chief Engineer's office

- clean up work areas and store all tools and materials prior to being relieved

ROUNDSMAN (Electromechanical Rooms)

Once at the beginning of the shift and every 4 hours thereafter, the roundsman will tour the facility and determine the operating status of all building systems and power plant equipment. On a log provided by the Chief Engineer, he will note his findings and take appropriate action when discrepancies are found. The remainder of his time, will be spent completing preventive maintenance procedures or other assigned work.

During his visits to the spaces, the roundsman will check the equipment and systems for:

- excessive noise/vibration
- overheating of system components
- broken belts/pulleys
- improper pressures/temperatures/levels
- inadequate guarding
- improper valving

- missing operating instructions
- torn/missing insulation
- broken/uncalibrated gauges
- open/broken access panels
- inadequate system identification
- water/oil/steam/air/fuel leaks
- use of power cord extensions
- broken mounts/pipes hangers/louvres, etc.

All paperwork generated during their respective shifts should be turned into the Chief Engineer, for review and follow up, by 8 a.m. the following morning. In addition to their regular assignments, it is expected that each of the operators will keep an ear peeled and a wary eye out for:

- out-of-date/unposted certificates
- blocked air intakes
- broken door locks/latches
- improper storage/clutter
- jagged edges and tripping hazards
- torn/missing walk-off mats
- missing fire extinguishers
- uncovered trenches/pits
- unsafe walkways, decks and ladders
- plugged equipment/floor drains
- burned out room/indicator lights
- unlabeled equipment/systems
- untagged valves/piping
- incorrect/missing directional signs
- grease/oil on floors
- missing ceiling tiles/floor grates
- improper storage of flammables
- excessive room temperatures
- lack of personal protective gear

Chapter 10

Loss Prevention/Recovery

Gather your marbles while ye may... and put them into a bag for safe keeping. In our case, that means the establishment of a loss prevention package comprised of an equipment conservation program and a boiler and machinery insurance policy. The key to successful retention of your marbles is in the strength of the packaging which contains them. Double bag them in heavy gauge plastic and a flooded room won't affect them. Put them in a paper sack and you'll lose them to a drip. The past few chapters constituted the gathering process; this one covers the bag.

Boiler and Machinery Insurance

However the policy is constructed, whether free standing or as an addendum to a fire insurance instrument, your fixed equipment can be insured against loss. Moreover, losses incurred as the result of equipment or system failure, such as rental fees or decreased production output, can also be covered. On the other hand, you may select (by choice or ignorance) to self-insure your operations.

When you purchase boiler and machinery coverage, you'll find it's packaged in many different ways. Like other insurances it has its deductibles, limits of liability, co-insurance provisions... etc., but the main concern should be the distinction made between actual cash value and the repair and replacement option. Under similar circumstances, when a claim is submitted for loss, an ACV policy will pay for the actual cash value of the damaged object after its value has been depreciated in accordance with a "schedule": whereas repair and replacement coverage assures the object will be made whole again regardless of the cost to the insurance carrier. Have the controller or risk manager contact the company's agent to determine what coverage is in force and make changes as appropriate. Once you're straight on the particulars, get back to work and pray you'll never have to submit a claim.

In the event that you do have insurance and suffer a problem, don't concern yourself with the cumbersome pages of legalize in your policy. All you need to know at that point is what triggers the recovery process and the telephone number of the guy (agent) who sees you through it. The "trigger" actuates when the occurrence meets the definition of an "accident" as defined by the policy. If you know when an accident occurs, you know whether to file a claim. This clause is typical of what you'll find written in most boiler and machinery insurance policies.

Definition

"Accident" shall mean a sudden and accidental breakdown of the object, or a part thereof, which manifests itself at the time of its occurrence by physical damage to the object that necessitates repair or replacement of the object or part thereof; but accident shall not mean:

a. depletion, deterioration, corrosion, or erosion of material;

b. wear and tear;

c. leakage at any valve, fitting, shaft seal, gland packing, joint or connection;

d. the breakdown of any vacuum tube, gas tube or brush;

e. the breakdown of any electronic computer or electronic data processing equipment;

f. the breakdown of any structure or foundation supporting the object or any part hereof, nor the functioning of any safety device or protective device

Investigation of Accidents

Depending on the extent of coverage and the magnitude of loss, the insurance carrier may dispatch an inspector and/or adjuster to your premises to substantiate the authenticity of your claim. The insurance representative will investigate the loss to a greater or lesser degree, questioning the involved/affected parties as to times, conditions... etc. and recommend actions to be taken in avoiding a recurrence. Based on the information gleaned from your records and people, and the terms of your boiler and machinery policy, a decision will be made as to the carrier's liability and funds will be disbursed to settle the claim. Needless to say, the better job you do of supplying the facts, the better chance you have of collecting. Forewarned being forearmed, the following outline is an actual guide utilized in the field by some insurance companies. Use it to prepare for their investigators. (B.I. stands for bodily injury).

**Figure 10-1
Loss Report Summary**

1. Summary: One sentence covering the substance of the incident.

2. Description: A "word picture" of conditions before the occurrence. The involved objects described in detail including,
   a. object identification
   b. all designating numbers
   c. size and capacity (service factor)
   d. protection devices (completely detailed)
   e. function of object (service in plant)
   f. operating schedule
   g. history of object (PM. operation, repairs... etc.)

3. Incident: Chronological report of event surrounding the incident
   a. events leading to discovery

(Continued)

		b. circumstances or known conditions
		c. action taken upon discovery

4. Damage and Repairs: Sketch: Detail physical damage and repairs in progress or proposed. (Include size, length, depth etc.)
   List recommended repairs in order of importance
   Describe which repairs are—
   - necessitated by incident
   - temporary and permanent
   - maintenance repairs
   - improvements

   Determine completion date for all repairs

5. Interruption To Production—State:

   a. product manufactured and normal quantity prior to interruption.
   b. percentage of B.I. (resultant interruption).
   c. means of B.I. expediting d. date plant expects to resume normal operations

   Include:
   a. possible extra handling of normal operation
   b. need for purchases (extraordinary) c. dependency on other plants
   d. downtime of other departments or equipment
   e. possible overtime

6. Cause: All factors responsible in a logical sequence of events. (Direct cause, subcause and contributory cause)

7A. Special Investigation   7B. Discussion
    Results of examination and lab tests

8. Conclusions and/or Recommendations:

(Continued)

> Recap main lessons brought out by incident
> Make practical detailed recommendations giving reason and benefits
> State briefly insured's reaction to proposed recommendations
>
> 9. Items of Interest:
> Comment on lack of cooperation on part of insured, negative management attitude, trade secrets pertinent to loss... etc.
> Human Element Comments:
> —apparent negligence
> —loss of life
> —confidential nature

Causes of Accidents

If normalcy can be described as a quintessential mixture of the materials, natural laws and human interplay required for its existence, then it stands to reason that accidents must occur when the amalgam is in some way deficient. Skeptical? Think back on any accident you've ever witnessed or was involved in. The likelihood is, if you look hard enough, the cause of every one of them can be traced to one of these:

- Material Failure
  (physical breakdown, design deficiency or chemical deterioration of a part, component or structure)

- Natural Phenomena
  (acts of nature and compliance with physics)

- Human Error
  (unsafe acts, negligence, omissions, incompetence, failure to perform, mismanagement, poor work practices, inadequate technical data, poor planning, inadequate supervision, erroneous instructions... GUESS WHO WINS?)

So what's the equation? One cause per accident? Not usually.

More likely there will be a direct cause such as an overload, one or more indirect causes such as defective conditions and a plethora of contributing causes such as inadequate standards and/or enforcement. The moral of the story?—NEVER ACCEPT THE MOST APPARENT CAUSE AS THE ONLY ONE OR YOU'LL NEVER CORRECT THE PROBLEM!

Pressure Vessel Failures

There are a lot of ways that your boilers, fired and unfired pressure vessels can bite the big one but time has shown that the majority of failures can be classified as bulging, rupture, collapse, cracking, overheating or leakage due to corrosion, erosion, wear or loss of joint or seam integrity. Reviewing the possible causal factors you'll find that most accidents can be *directly* attributed to:

- abnormal pressure
- abnormal temperature
- structural fatigue
- ignition of gasses/vapors
- inadequate support
- poor construction
- inadequate clearance
- physical weakening
- faulty
- design external mechanical damage

Indirectly traced to:
- misapplication
- abnormal loading
- normal wear
- normal aging
- faulty workmanship
- failure of controls
- failure of safeties
- improper assembly
- failure to protect
- foreign substances
- thermal shock

- faulty installation
- inadequate clearance

or <u>contributed to</u> by:
- nonperformance of maintenance
- noncompliance with recommendations
- operating in poor condition
- lack of operator attention
- improper operation/maintenance
- lack of training/supervision
- exceeding life expectancies of unit

Electromechanical Failures

Like pressure vessels, mechanical and electrical devices have their own peculiar ways of biting the dust. As a general rule, (at least from an inspector's view point), they fail by way of deformation, explosion, cracking, mechanical breakdown, roasting of insulation, weld fractures, electrical arcing, service Interruptions, or introduction of a foreign body. For these devices, the possible causal factors most accidents can be *directly* traced to are:

- abnormal expansion or contraction
- abnormal pressure load or voltage
- abnormal speed or speed fluctuation
- abnormal temperature (burning, roasting, single phase, friction, etc.
- abnormal vibration
- binding, seizing, rubbing, abrading, scoring, etc.
- electrical defect (short circuit, ground, contact or installation)
- flood, wind, rain, water, moisture, ice, etc.
- foreign source (oil, dust, dirt, object, animal or other substance)
- lightning
- mechanical defect
- mechanically unsuited for particular application
- not securely or properly fastened or aligned
- electrically unsuited for particular application
- fatigue or stress
- embrittlement

indirectly attributed to are:
- accelerated deterioration, abnormal loading
- misapplication, etc.
- lack of control or adequate protective devices
- defective material
- excessive wear and tear
- failure or malfunctioning of any protective or control device
- foreign substance (metal, dust, dirt or other substance)
- loss of or deficiency of lubrication or cooling medium
- normal deterioration—age or service
- water, moisture, rain, Freon, ammonia, oil, etc.
- stress concentration
- insulation (defective, improper type or improperly applied)
- poor workmanship or improper assembly
- shock (thermal or otherwise)
- material unsuitable for the service

or contributed to by are:
- failure to carry out recommendations
- failure to perform routine maintenance
- failure to remove from service when poor condition is known
- inadequate maintenance or supervision to prevent occurrence
- lack of operator or insufficient training
- misoperation of object
- failure to test or dismantle
- inspection—inadequate or incomplete
- manufacturing defect or design
- life expectancy for type of service
- no previous inspection

Boiler Control Failures

Why am I only addressing the controls on this particular piece of equipment? Because chillers shut down, but boilers blow up! Besides, more boilers are housed in more facilities than all the rest of the equipment we discussed combined. Moreover, as important as these devices are, they are frequently taken for granted, often with disastrous results. The most common boiler failure is from overheating due to operating the unit while it is suffering from a "low water" condition (water level in the boiler is below the normal operating level). Truth is, this problem was what led to the development of the automatic controls in the first place.

The two main controls I'm referring to are the feed water regulator, which causes water to be added to the boiler during operation to maintain the normal operating level and the low water fuel cut-off, which shuts the fuel supply off to the burner when the boiler water level drops below a predetermined, safe level. Each of these devices comes in a variety of types and actions. The imperative here isn't that you learn the particular way they work; better you understand why they fail.

No matter how good the design and construction of a control may be, its reliability depends on the maintenance and testing it receives. Failures fall into four categories: (1) installation, (2) mechanical, (3) electrical, or (4) human. For example:

1. *Installation*: (Mercury switches cannot make or break contact at the proper point if the control is not mounted in a level position.)

2. *Mechanical*: (A float in a water-control float bowl cannot rise and drop as it was designed to do if the float chamber is blocked by sediment.)

3. *Electrical*: (Switch performance in breaking the low-water cutout circuit fails when circuits are grounded in various places.)

4. *Human*: (An operator, either willfully or through lack of training, closes a main water feed valve or fails to prove that there is water in the boiler before starting the firing equipment.)

Both of these important devices should be tested frequently in accordance with the boiler and control manufacturer's recommendations and as required by your inspector. When testing controls on boilers in operation, the following precautions should be taken:

- Slow drain tests by shutting of feed supply and blowing down the boilers should not be made when operating at high capacity.

- When actual working tests of low-water fuel cutoffs are to be made and interruption of steam supply may affect plant production, make sure all departments concerned are notified that such tests are to be conducted.

- Open blowdown valves should never be left unattended. Make certain all blowdown valves are closed tight upon completion of tests.

- Boilers should never be blown down below the minimum safe water level. Some water should always be visible in the gage glass.

- When making actual working tests of low-water fuel cutoffs, observe proper safety precautions in restoring the fuel supply.

- Make certain that the water level is restored to normal and the burner equipment is operating properly before leaving the boiler room.

What's a slow drain test?—Basically, there are two ways of testing these controls. One is by the quick drain method and the other, the slow drain method. My recommendation? On low pressure steam boilers, controls should be quick drain tested at least weekly and slow drain tested quarterly. On high pressure steam boilers, they should be quick drain tested daily and slow drain tested monthly. Here's the procedure for each method as exacted on a low water cut-off.

Quick Drain Test—The quick drain test is applicable to low-water fuel cutoff and alarm devices having actuating elements (float or electrodes) located in a drainable chamber external to the boiler shell. The test consists of blowing down the chamber at a time when the burner is operating. If the device is functional, it should cause the burner to shut off and the alarm to sound. It will also flush the chamber and connections of accumulated sediment. Where dual cutoffs are installed each device should be tested independently of the other.

A quick drain test as described above is not applicable to so-called "built-in" low water cutoff or alarm devices having the actuating elements (float or electrodes) extending inside the boiler shell. Also, such a test is not practical on an operating hot water heating boiler. Some devices are so designed that a modified or partial test can be made without lowering the water level by manipulating the float mechanism or otherwise manually interrupting the electric circuit. This will test the circuitry but will not necessarily indicate the condition of the actuating elements.

Slow Drain Test—The slow drain test simulates a gradually developing low-water condition and is applicable to all types of low-water fuel cutoff and alarm devices on steam boilers. The test should be made with the burner in operation. Shut off the condensate return and feed supply so that the boiler will not receive any replacement water, and permit the water level to drop. The test may be expedited by opening the blowdown valves. The gage glass should be closely watched and the water level noted at the moment of cut off and/or sounding of the alarm. If the device fails to function and shut off the valve at the proper level, close the blowdown valves immediately and restore the water level to normal. Do not operate the boiler unattended until the cause of malfunction has been corrected. Each cutoff device should be tested independently of the other.

Boiler Explosion

Aside from your Chief Engineer, the most likely thing to explode in your power plant will be your boiler. But due to the many advances made over the years in the boiler industry, under the watchful eye of the A.S.M.E. (American Society of Mechanical Engineers) and thanks to the continual surveillance of the insurance industry and jurisdictional inspection agencies, it won't result from an overpressure condition but rather from a problem in the fuel trane. Most so-called boiler explosions are actually furnace explosions! Although they can and still do infrequently occur. The main causes? Sure...

- Insufficient purge before lighting off
- Human error
- Faulty low water cut off
- Faulty feedwater regulator
- Unbalanced air/fuel ratio
- Faulty fuel supply system
- Loss of furnace draft
- Faulty pilot ignitors
- False sense of security with subsequent failure of automatic controls

- Failure to test safety valves
- Failure to maintain auxiliaries
- Tampering with controls
- Lack of preventive maintenance
- Scale formation
- Internal corrosion (poor water treatment)
- Erosion (mechanical effect)
- Low water

CHAPTER 11
EQUIPMENT CONSERVATION

We've spent tens of hours discussing hundreds of issues relevant to the care and repair of all this stuff. If I never see another piece of plant equipment, it will be too soon! But before we wrap this up, I thought a reality check might be in order. The contractors are looking to turn everything over to us by the end of the month. Let's resurvey this engineer's dream from a biased perspective, to make sure we've covered all the bases. If we're lucky, maybe we can forego some future nightmares.

Evaluating The Operation

As we walk through the facility preparing our punch list, we should take a jaundiced view, aiming our sights at the operational continuity of the power plant. In manufacturing, system disruptions may result in decreased productivity—in business, declines in sales and in health care environments... even death. It's good to insure against losses, better to prepare for them but best to completely avoid them. How do we go about that? By simply asking the right questions up front and acting on the answers.

In prize fights ending in a knock out, I'm sure you'll agree that it isn't always the massive blow that concludes every contest. Most often the KO happens as the result of accumulated pummeling received over a number of rounds. But infrequently, a freak jab from out of no where has taken out many a worthy opponent. Call it the jab from Jersey, the punch from Panama or the hit from hell... it's quick, it's mean and it's over! In the power house, this type of hit comes from "weak link" failures such as the loss of the only hot water converter you have for supplying your domestic water tanks. Are you starting to get the picture? Good. Then let's start asking some questions—

- How reliable is the water supply? Electric? Gas? Is there a second source of feed available? What equipment relies on the supply for its cooling and/or operation? Are there restoration plans in effect, should the normal supply be interrupted or otherwise compromised?

- Does the plant have an appropriate equipment inventory? Is there adequate redundancy? Are there any one-of-a-kind devices? What about units of foreign manufacture? How long would repairs take? Are replacement parts/units readily available? Is the equipment properly located—in acceptable proximity of other/support equipment/auxiliaries?

- How functional is your repair capability? Do you stock adequate spares? Are your vendors reputable/dependable? What about lead times (including removal/reinstallation)? Can most repairs be done in-house? Is the equipment easily accessible? What's the worst case scenario for each system? Will any custom building be required? Have you given any thought to obsolescence?

- Are policies/procedures in place to assure continued operation? Do they spell out the actions to be taken? Who is responsible and what authority has he/she to act? Do they cover operator training? Inspection types/frequencies? Testing and maintenance? Recordkeeping?

Trouble Shooting System Design

Obviously, at this juncture, you're not in a position to reinvent your wheels but most certainly you should make yourself aware as to why they wobble. At best, knowing your equipment foibles enables you to predict future outages; at least, it allows you the luxury of scheduling shutdowns in order to avoid the unscheduled variety. If you're fortunate enough to be calling the shots on new equipment purchases, opt for proven reliability, reasonable expandability, maximum flexibility, ultimate safety and installation as close to the load as possible. The foibles?

- untested performance
- improperly located
- undersized for load being handled
- improper clearances/tolerances
- faulty operating/safety controls
- inability to meter loading
- lack of instrumentation
- uncoordinated protective devices
- lack of access for maintenance
- history of abusive operation
- limited usage/adaptability
- incompetent operators
- no spare parts available
- not to mention excessive vibration, leakage, problems with overheating, nuisance tripping, deterioration, air infiltration... etc.... etc.... etc.

Contingency Planning

Even with your eagle eye for detail, your hawkish attack of plant problems, and your mother hen devotion to your operations, you can still be given the bird by your systems or the utilities supplying them. Unexpected outages and shutdowns, though (hopefully) not routine, are nonetheless a fact of life in our aviaries. They arrive on the wings of your 13th operator (Mr. Murphy) and regardless of the comprehensiveness of your programs, there's little you can do to avoid them. Gas lines rupture components fail—power lines go down. Hey... stuff happens! And you aren't always able to control it. But when your plant gets dumped on, there's no need to get your feathers ruffled, as long as you've installed a plan for circumventing the problem.

In the event of a system shutdown or cessation of utilities, your operators must know exactly what steps to take in order to normalize the situation. They must act quickly and correctly to assure continuity of operation during dire circumstances. A simple set of emergency instructions, such as these, should be prominently posted in the engineers office. They will provide the immediate (down and dirty) information the operator needs to assure operating continuity and get the right people involved in the resolution of the problem.

Figure 11-1

PLANT MAINTENANCE		
POLICY TITLE:	Emergency Instructions	POLICY NUMBER: 3-07.11.1.0
DATE OF ORIGIN: 11/19/92 REVIEWED & REVISED	FOR DIVISION USE FOR DEPARTMENT USE X	PAGE 1 OF 5

I. PURPOSE:
 	
    To insure the continuous operation of Physical Plant systems and equipment during crisis situations.
 	
II. ORIGINATOR:
    Director of Facilities Management
 	
III. SCOPE:
    Plant Management, Plant Maintenance		

APPROVED:		

                (Signature)   (Title)   (Date)

(Continued)

Equipment Conservation

3-07.11.1.0

IV. TEXT

Prior to the implementation of the below guidelines, the Director of Facilities Management and/or the Chief Engineer must first be notified.

The following list denotes the action that should be taken as the result of equipment failure or system interruption.

A. Boilers

1. <u>Loss of One Boiler</u>
   a. Immediately transfer load to idle boiler.
   b. Notify all affected departments that an interruption of service has occurred, indicating approximately when service will continue.

2. <u>Loss of Both Boilers or Steam System</u>
   a. All parameters of the disaster manual covered under the boiler section go into effect.
   b. Notify all affected departments of the situation status.
   c. If repairable, contact Cleaver Brooks, Inc. at 1-800-535-4652.

B. Chillers

1. <u>Loss of One Chiller</u>
   a. Immediately transfer the load to the idle chiller
   b. Notify Data Processing, and Departments of the situation and have them monitor the room temperatures for their critical equipment.

2. <u>Loss of Both Chillers</u>
   a. Contact the Switchboard to make a general announcement over the public address system that the air conditioning system is down.
   b. Notify _____ Refrigeration at _____ for service

(Continued)

C. Vacuum System

1. <u>Loss of One Vacuum Pump</u> - Transfer to the alternate pump.
2. <u>Loss of Entire System</u>
   a. Notify XXX Engineering at (XXX) XXX-XXXX

D. Compressed Air System

1. <u>Loss of One Compressor</u> - Transfer to the alternate compressor. Call in Compressor Engineering at (XXX) XXX-XXXX
2. <u>Loss of Entire System</u>
   a. Notify switchboard to make a general announcement of system outage.
   b. Notify

E. Plumbing System Failure

1. <u>System Component Loss</u> - Refer to Valve List for appropriate isolation of the system until repairs are made.
2. <u>Major System Loss</u> -
   a. Shut off all water supply to the building at the main inlet valve. Contact the XXXXX Water Department.
   b. In either case, notify the affected departments indicating approximate repair completion times.
   c. All parameters of the disaster manual's water section go into effect.

F. Electrical Power Interruption

1. <u>System Component Loss</u> - Assure that the branch circuit of the emergency generator is activated for the section in question.
2. <u>Total System Loss</u>
   a. Immediately check the emergency generator system referring to the emergency generator operating manual.

(Continued)

b. Notify the switchboard to make a general announcement of the loss indicting that only the emergency circuits are energized.
c. In the event that the generator is inoperable contact XXXXX at (XXX) XXX-XXXX
d. Contact the Electric Power Company to determine the status and duration of the outage.

G. Communication System

1. Refer to Communication section of the Disaster Manual.
2. Contact XXXXX at (XXX) XXX-XXXX for repair assistance.
3. Apprise all affected departments of their communication alternatives.

H. Natural Gas

1. Shut off main supply valves.
2. Transfer boilers to oil-fired operation.
3. Notify Switchboard to make a general announcement.
4. Contact affected departments with situation assessment.
5. Contact XXXXX at (XXX) XXX-XXXX for service.

I. Fire Warning System

1. Notify XXXX Fire Department of any system problems.
2. Have Switchboard make general announcement of system status.

J. Elevators

1. Insure that all elevator passengers are safely removed from the cars.
2. Shut down elevators until repairs have been effected.
3. Contact XXXX at (XXX) XXX-XXXX for services.

Should that be the extent of the operator's involvement, such initial action is tantamount to pulling an alarm and attempting to confront a conflagration with a hand held extinguisher until the fire department arrives. Having notified the proper agencies, more elaborate procedures/precautions might be taken, as outlined in the power house policy/procedure manual. Containing a fire, for example, might call for the orderly killing of electrical power to affected areas, compartmentalization of the structure or activation of emergency backups. Once the operator has gained a modicum of control over the situation, these more specific instructions can than be acted upon to further alleviate the problem. Training of personnel in the proper implementation of these procedures and running simulated drills will clip Murphy's wings. And it's important you do both—if you leave him a wing, he'll have you flying around in circles.

The Concept of Maintenance

Without delving into the particulars of its variations, equipment maintenance can be defined simply as... "the manipulation of devices required to keep them in good working condition." In short, all maintenance activity falls into four basic categories—cleaning, calibration, lubrication and parts replacement.

1) Cleaning
   Moisture, rust, scale and accumulations of
   dirt, dust and debris are major
   contributors to overheating of cooling
   systems, arcing in electrical circuits,
   decreased efficiency of heat, transfer
   surfaces and sticking of moving parts
   just to name a few.

2) Calibration
   Improper control settings, inaccurate
   readings and uncoordinated operating and
   safety interlocks result in premature
   start up, shutdown, inappropriate cycling,
   system interruptions, outages and
   production losses

3) Lubrication
   Poor lubrication choices, lubrication techniques and frequencies lead to ceasing of moving parts, overheating of metal surfaces from lack of cooling and infiltration of dust and dirt into internal workings.

4) Parts Replacement
   Worn out consumables such as cables, belts, filters, hoses etc., cause undue strain on the equipment during operation, misalignment of components, loosening of parts inadequate distribution of fluids and inefficient transmission of power.

The engineers' credo should be keep it clean, keep it tight, keep it oiled, keep it running. On boilers, fired and unfired pressure vessels maintenance efforts should be focused on the cleaning of fire and water sides, testing of controls and overpressure protection devices, calibration of combustion safeguards and renewal of seals and gaskets. Maintenance of electrical devices should consist of checking wiring insulation, insuring proper closure of switching mechanisms, calibration of meters/instrumentation, physical tightening of connections, integrity of grounding and protective devices, and cleanliness of busses and component parts. Mechanical maintenance calls for bearing and gear replacements, cleaning and lubrication of moving parts, cleaning of filters and strainers, calibration of instrumentation and testing of operating controls and safety interlocks. In all cases, maintenance activities should begin with the recommendations outlined in the manufacturers operating/instruction manual.

Nondestructive Testing

NDE is the science of detecting minute surface cracking and/or subsurface imperfections in component parts of electromechanical devices and pressure vessels without actually modifying or otherwise affecting their structural integrity. The detection of material flaws without the need for tearing equipment apart and/or damaging it in the process is a quick, cost effective means for uncovering potentially serious problems in their developmental stages which would otherwise go unnoticed until culminating in a failure of the part or unit.

The process incorporates sophisticated apparatus in the performance of proven analytical techniques for diagnosing such maladies as cracking or thinning of pressure vessel walls, failure of tube walls in coils, loose electrical connections, discontinuities in weldings, and heat loss from inadequate insulation. Types of examinations include:

Dye Penetrant Testing

Works in the same manner as the fluorescent-penetrant method, but does not require black light. A red dye is applied to the surface, followed by a white developer which dries on contact. Any dye which has seeped into cracks in the metal will stain the white coating and outline the flaws in red.

Eddy Current Testing

When a conductor is placed in a primary magnetic field, it generates a secondary magnetic field. If the primary magnetic field is created by an electromagnet using alternating current, a secondary alternating magnetic field is created in and around the conductor. This induces an alternating voltage within the conductor. The resulting current follows a circular path and is termed an eddy current.

The eddy current test method is most commonly used for checking variations in wall thicknesses, longitudinal seams and cracks, pits, porosity, metal analysis, lack of bond, and thickness coating. It is used on items such as tubing, bar stock, ball bearings, wires, welded parts, and flat or sheet stock.

Four significant properties of the conductor (or test object) may be measured: electrical conductivity, magnetic permeability, density, and geometry (including surface and subsurface discontinuities). Further, the distance from the primary magnet and the conductor, or an irregularity within the conductor, can be accurately measured. If the intervening space contains a nonconductor, the thickness of that material is measured.

Eddy-current testing can be used in either ferrous or nonferrous piping and tubing to determine thinned sections and search for surface or slightly subsurface faults. Current is passed through a coil surrounding, or inside, the pipe and any irregularities or different thicknesses will produce changes in coil impedance and be indicated on a meter, recorder or oscilloscope. Unlike the above methods, this is a one-step operation—quick and simple.

Fluorescent-Penetrant Inspection—Will indicate surface defects on both magnetic and nonmagnetic materials (such as aluminum). A fluorescent liquid is applied to the area in question, followed by a developer. When examined under a black light, any discontinuities in the metal will fluoresce.

Magnetic Particle Flaw Detection—Can be used on ferromagnetic materials for detection of surface or slightly subsurface cracks. An area of metal is magnetized and magnetic particles are applied to the surface. These particles will tend to collect around irregularities where lines of force are strongest, thus clearly outlining any cracks.

Radiography

There are many different radiographic techniques such as fluoroscopy, radiography, three-dimensional and color radiography, flash or high-speed radiography, vidicon radiography, etc. However, the common workhorse for maintenance applications is X-ray and gamma-ray film radiography. This involves exposing film to X-rays or gamma-rays that have travelled through the material or equipment to be tested.

Radiographs are used to detect internal defects such as cracks, voids, lack in bonding, laminations, and porosity. Typical applications include testing of sheets, tubes, rods, forgings, castings, welds, seams, lamination, honeycomb, brazing, assemblies, and pressure vessels.

Although radiography requires rigid safety controls and is expensive, it offers important advantages: a permanent record is provided on film; the detection of internal irregularities in density or geometry is possible with high sensitivity; and small sources of radiation can be placed in small cavities for interior shots.

Radiography is a highly efficient means of examining for subsurface flaws in either magnetic or nonmagnetic materials. The surface is exposed to short wave radiation which, after passing through the material, leaves a shadowgraph image on X-ray film attached to the rear of the plate. Experienced technicians are needed to interpret accurately the results of such tests. Repeated exposure to radiation is a health hazard, so tests areas should be marked off and protection provided for the inspection team.

Thermography

The use of infrared pictures or thermography, is evolving as a promising tool for maintenance. This heat detection technique has been used by utility companies for several years. The chemical industry is now applying infrared scanning (by television camera) to show live pictures of the temperature differences on the surfaces of equipment. A typical picture will show variations in shading, from the darker, low-temperature surfaces to very light, high-temperature surfaces. The distance of the camera from the object may be as much as several hundred feet with no effect on the highly accurate readings, and using a known heat source for calibration permits accuracy to within one-half degree. The temperature may range from –22°F (330°C) to more than 3,600°F (1,982°C).

Principal applications have been in detecting "hot spots" in lines, leaking relief valves, furnaces, induction coils, reactors, process valves, heat exchangers, switchgear, substation lines, and disconnects. In switchgear, contacts and busbars may be examined without power interruption for hot spots in areas where trouble is developing.

The necessary equipment is expensive: color units rent for about $2,000 to $8,000/month. Visual access is required to get a picture.

Although there is still limited experience in applying the technique—there are only a limited number of newer units in use—thermography is growing in use and application. When vital equipment cannot be shut down, the investment is small in terms of the information gained. Thermal Imaging (INFRARED) used for the detection of "hot spots" such as might result from defective refractory. A camera surveys the affected area for infrared light and transmits an image to a screen. Hot areas will appear brighter than cooler areas because they give off more of this infrared radiation.

Ultrasonics

Many ultrasonic techniques are based on two primary means of transmission. These are the "pulse-echo" and the "dual transducer," or "through-transmission," methods. The pulse-echo method alternately transmits sound waves and then receives the reflection. Variances in reflection time indicate a discontinuity in the tested material. The through-transmission method transmits and receives at the same time, with transducers on either side of the tested material. Any interruption in the sound wave will appear on the readout device, possibly indicating a discontinuity in the test material.

Ultrasonic techniques can detect unbonded areas, delaminations, cracks, voids, microfissures, inclusions, corrosion, pitting, forging and welding flaws, and lack of adhesion.

Some of the advantages of ultrasonics include: dependability, ease of operation, deep penetration, prompt results, high accuracy and sensitivity, automated testing, a variety of readouts, and the need to utilize only one surface. Ultrasonic Testing detects both surface and deeply hidden discontinuities in material. Electrically timed sound waves are transmitted through the metal and are reflected back off the rear surface. should a wave hit a break in the metal, however, it will be reflected back sooner than other non-interrupted waves, as it has a shorter distance to travel to the wave source. (The atmosphere on the other side of the plate is treated as an infinitely large "crack.") This process is also ideally suited to measuring wall thicknesses where wastage due to corrosion or erosion is suspected.

Vibration Analysis

Vibration is always present when any machine is converting energy into useful work. In simple terms, it is the motion of any member or part moving back and forth from its position of rest. The total distance of movement is known as the peak-to-peak displacement or amplitude. The frequency is the number of cycles of this movement over a given period of time. A third characteristic is known as the phase of vibration. The fourth characteristic is the velocity or speed with which the position of the part is changing.

When a machine is operating normally and according to design, it will produce a certain vibration "signature." Any deviations from this normal pattern of vibration is a signal that the machine may require maintenance. Principal causes of abnormal vibration are: (a) imbalance, (b) misalignment, (c) worn bearings, belts, or gears, (d) aerodynamic or hydraulic forces, and (e) reciprocating forces.

A vibration analyzer is capable of measuring the displacement (amplitude), the frequency, and the phase of vibration. When vibration occurs at several frequencies, the analyzer must be able to separate one frequency from another in order to measure the characteristics of each.

Vibration probes can be permanently installed on operating equipment and vibration characteristics are monitored continuously, permitting diagnosis of operating problems as needed.

Vibration analysis can reduce costs by in-place balancing, early detection of trouble, elimination of damaging failure, evaluation of new and repaired equipment as to quality and proper installation, and by providing specific knowledge of what replacements and repairs should be made. It is a relatively inexpensive technique, though, while changes in vibration patterns are quickly detected, it may require sophisticated analysis to determine what caused the change.

CHAPTER 12

ENERGY CONSUMPTION

The project is quickly drawing to a close and my creative juices have all but been spent. Unlike that never ending battery commercial, my energy reserves aren't limitless. But don't let that rabbit fool you; he's just going... and going... and going to the bank. In the real world, if you don't monitor energy consumption you can bank on your rabbit dying, then you can watch your operations go... go... go down the tubes. I've got enough drum left to pound out one last message. Move over, Bugs!

The Building Envelope

Like wood nymphs, sprites and poltergeists, "energy" is elusive and comes in many forms. Pinning it down to a singular essence would be like grouping all imaginary beings as the same. The reality is... just as leprechauns are elves, but all elves are not leprechauns; electricity is a form of energy but not all energy is electric. And no where is that truth more evident than in the energy intensive power plant. There is potential energy in our fuel supplies, kinetic energy in our pipes and expended energy in the heat we're constantly transferring, supplying or dissipating with our machines and devices.

After recognizing energy's many faces (chemical, mechanical, thermal... etc.), it's the monitoring and control of that "heat" that gives the building engineer his most pause. A building's shell (or envelope) provides a barrier between the unpredictable atmosphere outside and the controlled environment within. Properly designed and utilized, this barrier, which includes all of the buildings structural components (doors windows, roof, walls... etc.), can selectively filter out or allow in certain amounts of fresh air, light, heat and humidity to temper the buildings interior as desired. A good envelope will:

- minimize heat losses/gains through proper insulation and control of sunlight

- maintain movement of water vapor and control condensation via good ventilation and humidity control

- prevent inappropriate air infiltration and escape through engineered openings

- provide adequate venting of interior gasses with mechanical exhaust systems

Equipment Operation/Maintenance

Open that envelope up and you'll probably find a stack of bills from your local energy suppliers. You purchase natural gas for your boilers, electricity for your motors and lights, water for process and consumption, fuel oil for your generators, and gasoline for your chicles and portapumps. If energy were a drug, you'd be a junkie. How much does it all cost? Do you care? Should you? What are you going to do to cut your dependency? Here are some general rules of thumb—followed by a few system specific suggestions:

(General)

- when replacing equipment, make sure it is properly sized for the load to which it will be connected

- if it's not being used, turn it off!

- when purchasing anew, buy the more energy efficient model

- lower outputs only to needed levels

Energy Consumption

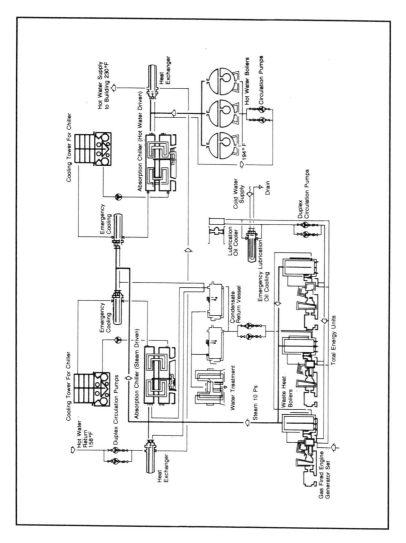

Figure 12-1

- avoid excessive demand charges by operating your equipment during "off peak" hours

- lower settings to reduce consumption (temperature, pressure, speed...)

- don't overload, but eliminate partial loading; run at full capacity

- keep all moving parts clean and well lubricated

- adjust linkages/alignments and replace worn parts

- automate manual controls; calibrate existing controls/devices

- schedule equipment for overhaul/shut downs; perform internal inspections

- utilize waste heat from processes to produce steam,

- preheat fluid or warm spaces

- insulate... insulate... insulate!

(Specific)

Boilers
- check furnace combustion efficiencies and make adjustments regularly

- replace inefficient burners with an automatic combustion control system suited to the fuel being burned

- lower boiler operating pressures and/or install a "summer" boiler for periods of lesser loading

- install turbulators, automatic flue dampers and insulation in gas passes as appropriate

- modify the system to eliminate distant branches; install separately controlled zones and/or localized heating/cooling equipment

- check and maintain steam traps and condensate return piping

Electric Motors
- don't rewind burned out motors; replace them with higher efficiency rated units

- oversized motors operate at reduced efficiencies; size the motor to the load

- equip oversize motors with variable speed drives and/or consider ganging smaller motors to meet varying system needs

- install load shedding, peak shaving and power factor correcting devices to reduce demand charges

- turn off equipment and de-energize circuits when not in use

HVAC
- reduce lighting levels and duration in air-conditioned spaces

- correct situations where areas are heated and cooled simultaneously

- experiment with raising humidity to compensate for lower temperatures

- utilize outside air to provide "free cooling" of interior spaces

- insulate air ducts, chilled/hot water lines and steam piping

- convert air handlers to (VAV) variable air volume type with variable speed blowers

- replace filters as required by system monitors

Kitchen Equipment
- avoid unnecessary preheating of cooking devices use only as high a temperature as is required

- improve and regularly maintain ventilation and exhaust systems

- install only energy efficient models such as infrared fryers, microwave ovens and specialized, food specific devices

- utilize waste heat generated in the kitchen for preheating process air and water

Miscellaneous Industrial
- set controls on refrigeration equipment only as low as necessary

- turn machines off when not in use; don't leave stand-by units energized unnecessarily

- keep nozzles, intakes, drains and orifices unclogged and free flowing

- install heat reclamation devices and eliminate hot water leaks/drips

- reduce water temperatures in hot water systems to acceptable sanitation levels

- turn off water heaters or circulating pumps when system isn't being used

General Conservation Measures

Reducing energy consumption is a lost art. It seems that even the best programs can get off track. Maybe we don't establish it as a high enough priority. Possibly we aren't providing the proper training for our personnel. Perhaps people just don't care when they aren't footing the bill. Do we take conservation measures only when we're afraid we're running out? I guess it all gets back to energy's elusive nature. Energy itself may be considered one of life's great mysteries but it doesn't take a shaman to figure out the consequences of over utilization of our limited resources.

Essentially, there are two approaches to reducing energy consumption: one is *pro-active*, where we take the bull by the horns and wrestle it to the ground; the second is *inactive*, where we do nothing and the bull eventually dies and falls down with a thud. Granted, the first option is a lot more work but the alternative... I mean... can you imagine the stink that would cause? Then there's the burial! Every plant should undergo a comprehensive audit, performed by a competent energy specialist which culminates in an established (enforced) conservation program. Until you can arrange that, here are some additional Heloises to get by on:

Daily
- switch off safety/security lights at daybreak

- utilize lighting/heating/air-conditioning only as required

- take advantage of daylighting situations in lieu of electric lighting

- preadjust thermostats for comfort cooling/heating

- set back thermostats later in the day prior to exiting

- keep doors/windows closed; make use of Venetian blinds

Weekends
- turn off all but safety/security lighting

- set back heating/cooling thermostats

- check settings and accuracy of HVAC 'controls

- de-energize water heaters and domestic water circulating pumps

Seasonal
- prior to heating and cooling seasons; thoroughly test, inspect, and adjust all associated equipment and controls in accordance with manufacturer's recommendations

- turn off heating in unoccupied spaces that won't be adversely affected by the lack of same

- in hot weather, take advantage of night air to precool the building

- thoroughly winterize buildings and equipment in accordance with the dictates of the energy management program guidelines

- adjust outdoor timers to accommodate the seasonal temperatures and hours of daylight

- cover or remove window air conditioners at the onset of the heating season

Construction Considerations

It's arguable that the following items should be addressed by whomever is charged with overseeing construction and/or renovation projects in your organization. But as you're the one providing the utility power it won't hurt to have a nodding acquaintance with the subject when they come to you for your advice. When they query you for your input; ask them if they planned for:

- efficient glazing systems; sufficient thickness/values of insulation in appropriate locations

- high pressure sodium lights for parking lots and loading docks

- daylighting versus electric lighting

- energy efficient fluorescent fixtures/ballasts

- timers/photoelectric controls and occupancy sensors

- building envelope considerations such as—thermal windows, adjustable shades, 'window overhangs, revolving doors, wind breaks... etc.

- location of support equipment in close proximity to the loads and areas served

- tapping into existing systems

- pipe/duct insulation, low flow shower heads, faucet controls, programmable thermostats, adjustable vents/dampers

- proper sizing of HVAC systems/components; use of heat pumps

- zoning of systems; energy efficient machinery; peak shaving; energy management systems... etc.

Figure 12-2

Energy Management Systems

Want to improve your stock when it comes time to resell your properties? Tell the prospective buyer they come equipped with computerized energy management systems. Without getting into the subjects of energy audits, life cycle costing and rate of return analysis, it's conceded by most that a well conceived energy management program will provide a good return on investment (ROI), with payback periods of one to three years as the acceptable norm. But before you go running off to purchase yours, it might pay you to hire an expert to walk you through the process. A good analyst (non-psychiatric) will be able to educate you to what's "state of the art" in the energy conservation field, analyze those parts of your systems you don't understand, estimate potential costs/savings, explain possible modifications and suggest computer hardware/software acquisitions. When you contract with one, make sure it's spelled out what specific services will be provided, what reports will be forthcoming (including audits) and how much your operation can reasonably expect to save over what period of time.

Once the audit (analysis) is out of the way, an energy management system can be chosen which best matches your established needs. Any EMS worth its chips should be able to:

- turn off lights in occupied areas
- maintain partial lighting before and after "public" hours
- schedule lighting operation by hour of day, and time-of-year
- turn off a water heater when appropriate
- eliminate hot water circulation when an area is unoccupied
- maintain HVAC system start-up and set-back schedules
- eliminate unnecessary HVAC use during unoccupied hours
- monitor/control space temperatures/humidity
- provide temperature/pressure read outs on selected equipment

Bear in mind that when you interrupt, modify or otherwise change your existing systems, you take the chance of affecting the productivity, health and comfort of the buildings machinery and occupants, as well as falling out of compliance with applicable codes/ standards.

Well guy (girl), it's been real! Let me know how you make out and don't hesitate to call me if you need my help. TTFN

Appendix A
RFQ Boiler Plate

Minority and Resident Participation

Subject	Page Number
Basis of Contract Award	CF-1
Trade Participation and Reporting	CF-2
Major Trades or Crafts	CF-3
Minority and Female Owned Business Form	CF-4
Trade Participation Form	CF-5
Affidavit	CF-6

Basis of Contract Award

In order to promote an intended goal to maximize the use of minority and female personnel, minority and female owned business and craft work performed by _____ County residents on these Projects, the County of _____ sets the following goals for Contract awards.

1. Minority/Women Participation:
   A. Twenty-five percent of the dollar amount of the bid total should be awarded to minority owned business(es). Five percent of the dollar amount of the bid total should be awarded to female owned business(es).
   B. Minority and female owned businesses must be identified on form CF-4, indicating the above goals.

2. Minority and Women Personnel and _____ County Residents:
   A. It is the intention of the County that worker hours be performed by minority and women personnel and by residents of the County. In this regard, the following percentages should be met on a craft-by-craft basis: 50% of worker hours are to be performed by residents of _____ County, 25% of worker hours are to be performed by minorities, and 10% of worker hours are to be performed by women.
   B. Personnel must be identified on form CF-5, indicating the above.
   It is the intention of the County that the bidder make every effort to reach the goals as outlined above. Following the bid opening and prior to the award of a contract to an acceptable bidder, the acceptable bidder will be required to submit executed copies of his/her contracts with these minority and women owned businesses listed on page CF-4.

CF-1

Trade Participation and Reporting

In accordance with this commitment, the Contractor must submit both the Payroll form and the Total Manhours work sheet on a bi-weekly basis during the course of construction (See sample attached). All Sub-contractors are to be listed on the Total Manhour sheet whether they are active or not. For the purpose of this report, the following group categories will be used:

A. Black American
B. Hispanic American (with origins from Puerto Rico, Mexico, Cuba, South or Central America, Spain, Portugal, or the Caribbean Islands regardless of race)
C. Native American (American Indian, Eskimo, Aleut, Native Hawaiian)
D. Asian-Pacific American (with origins from Japan, China, the Philippines, Vietnam, Korea, Samoa, Guam, the U.S. Trust Territories of the Pacific, Northern Marianans, Laos, Cambodia, Taiwan, or the Indian Subcontinent)
E. An individual of any nationality, race or religious group currently qualified for protected class status by determination of the U.S. Supreme Court.
F. White persons of Indo European descent, including Pakistani and East Indian)

A minority group member is an individual who is a citizen of the United States, and is identified in one of the aforementioned categories A through E.

The term "Minority Business Enterprises" (MBE) means a business at least 51 percent (51%) of which is owned and controlled by minority group members. The minority ownership must demonstrate actual day-to-day participation and control.

CF-2

Major Trades or Crafts

Asbestos workers
Boiler Makers
Bricklayers
Carpenters
Cement Masons
Electricians
Elevator Construction
Glaziers
Machinists
Machinery Movers
Ornamental Iron Workers
Lathers
Operating Engineers
Painters
Pipe Driver Mechanics
Pipe Fitters/Steam Fitters
Plasterers
Plumbers
Roofers
Sheet Metal Workers
Sprinkler Fitters
Technical Engineers
Tuck Pointers

For approval of other trades for consideration, written approval should be requested from the Purchasing Agent of _____ County.

CF-3

Appendix A—RFQ Boiler Plate

Minority and Female Owned Businesses
(to be included with Proposal)

The following information must be supplied by the Contractor for the purposes of evaluating the Contractor's efforts. This list does not abrogate the County's right to approve all Sub-contractors and vendors proposed for use on the contract.

Type of Work	Business Name Address & Phone #	Ownership/ Gender	Dollar Amount of Contract

Total Dollar Amount Female Owned Businesss .... $ _____
Total Dollar Amount Minority Owned Business .... $ _____
Total Dollar Amount Minority and Female Owned Business ...... $ _____
Percent (%) of Total Contract .... $ _____

CF-4

**County Residents, minority Personnel Participation Form
(to be included with proposal)**

Anticipated levels of minority participation on a worker-hours basis to be expressed as percentages must be supplied for each trade, or craft, whether attributable to the Contractors work force or any subcontractor, which will be active on these projects.

Position	% By County Residents	% By Minority	% By Female
_____	_____	_____	_____
_____	_____	_____	_____
_____	_____	_____	_____
_____	_____	_____	_____
_____	_____	_____	_____
_____	_____	_____	_____
_____	_____	_____	_____
_____	_____	_____	_____
_____	_____	_____	_____
_____	_____	_____	_____
_____	_____	_____	_____
_____	_____	_____	_____
_____	_____	_____	_____
_____	_____	_____	_____
_____	_____	_____	_____
_____	_____	_____	_____
_____	_____	_____	_____

Note: If additional spaces are needed, the enclosed form may be copied as many times as needed

CF-5

Appendix A—RFQ Boiler Plate

Affidavit

I, _____
            (Individual's Name)

the _____
            (Individual's Title)

of _____
           (Company's Name)

bidder on the entitled project being first duly sworn on oath, deposes and says that the information supplied by the bidder for the purposes of evaluating the contractor's use of minority and women personnel and the use of minority and female owned businesses in the performance of this contract, is true and correct.

_____
Signature

Sworn And Subscribed To Before Me This

_____ day of _____, 19___

_____
Notary Public

(NOTARY SEAL)

CF-6

Office Of The Purchasing Agent

County

President, Board of Commissioners   Purchasing Agent

Index

Description	Page
Minority and County Resident Participation	PF-1/5
Certificate of Qualification	PF-6/7
Sole Proprietor	PF-8
Partnership	PF-9
Corporation	PF-10

Minority And County Resident Participation

Basis Of Contract Award

In order to promote an intended goal to maximize the use of minority and female personnel, minority and female owned business and craft work performed by County resident on these Projects, the County _____ sets the following goals for contract awards.

1. Minority/Women Participation

    A. Twenty-five (25) percent, of the dollar amount of the bid total should be awarded to minority owned business(es). Five (5) per cent of the dollar amount of the bid total should be awarded to female owned business(es).

    B. Minority and female owned business must be identified on form CF-4, indicating the above goals.

2. Minority and Women Personnel and _____ County Residents:

    It is the intention of the County _____ that worker hours be performed by minority and women personnel and by residents of the County _____. In this regard, the following percentages should be met of a craft-by-craft basis: 50% of worker hours are to be performed by residents of _____ County, 25% of worker hours are to be performed by minorities, and 18% of worker hours are to be performed by women.

    Personnel must be identified on form CF-5, indicating the above.

It is the intention of the County _____ that the bidder make every effort to reach the goals as outlined above. Following the

PF-1

bid opening and prior to the award of a contract to an acceptable bidder, the acceptable bidder will be required to submit executed copies of his/her contracts with those minority and women owned businesses listed on Page CF-4.

Trade Participation And Reporting
In accordance with this commitment, the Contractor must submit both the Payroll form and the Total Manhours work sheet on a bi-weekly basis during the course of construction (see sample attached). All sub-contractors are to be listed on the Total Manhour sheet whether they are active or not. For the purpose of this report, the following group categories will be used:

A. African American;

B. Hispanic American (with origins from Puerto Rico, Mexico, Cuba, South or Central America, Spain, Portugal, or the Caribbean Island regardless of race);

C. Native American (American Indian, Eskimo, Aleut, Native Hawaiian);

D. Asian-American (with origins from Japan, China, the Philippines, Vietnam, Korea, Samoa, Guam, the U.S. Trust Territories of the Pacific, Northern Marianans, Laos, Cambodia Taiwan, or the Indian subcontinent):

E. An individual or any nationality, race or religious group currently qualified for protected class status by determination of the U.S. Supreme Court.

F. White (persons of Indo European descent, including Pakistani and East Indian).

A minority group member is an individual who is a citizen of the United State, and is identified in one of the aforementioned categories A through E.

PF-2

The term "Minority Business Enterprises" (MBE) means a business at least 51 percent (51%) of which is owned and controlled by minority group members. The minority ownership must demonstrate actual day-to-day participation and control.

Major Trade or Crafts

Asbestos Workers	Boiler Makers	Bricklayers
Carpenters	Cement Masons	Electricians
Elevator Construction	Glazier	Machinists
Machinery Movers	Ornamental Iron Workers	Lathers
Operating Engineers	Painters	Pipe Driver
Mechanics	Pipe Fitters/Steam Fitters	Plasterers
Plumbers	Roofers	Metal Workers
Sprinkler Fitters	Technical Engineers	Tuck Pointers

For Approval of Other Trades For Consideration, Written Approval Should Be Requested From The Purchasing Agent of _____ County.

PF-3

Minority And Female Owned Businesses (To Be Included With Proposal).

The following information must be supplied by the Contractor for the purposes of evaluating the Contractor's effort. This list does not abrogate the County's right to approve all sub-contractors and vendors proposed for use on the contract.

Type Of Work	Business Name Address & Telephone Number	Ownership/ Gender	Dollar Amount Contract
_____	_____	_____	_____
_____	_____	_____	_____
_____	_____	_____	_____
_____	_____	_____	_____
_____	_____	_____	_____
_____	_____	_____	_____
_____	_____	_____	_____
_____	_____	_____	_____
_____	_____	_____	_____
_____	_____	_____	_____

Total Dollar Amount
   Female Owned Business   _____

Total Dollar Amount
   Minority Owned Business   _____

Total Dollar Amount
   Minority and Female Owned Business   _____

Percent (%) of Total Contract   _____

PF-4

County Residents, Minority Personnel Participation Form (*To be Included With Proposal:

Anticipated levels of minority participation on a worker-hours basis to be expressed as percentages must be supplied for each trade, or craft, whether attributable to the Contractors work force or any subcontractor, which will be active on these projects.

Position	Percent By County Residents	Percent By Minority	Percent By Female
_____	_____	_____	_____
_____	_____	_____	_____
_____	_____	_____	_____
_____	_____	_____	_____
_____	_____	_____	_____
_____	_____	_____	_____
_____	_____	_____	_____
_____	_____	_____	_____
_____	_____	_____	_____
_____	_____	_____	_____
_____	_____	_____	_____
_____	_____	_____	_____

Note: If Addition Spaces Are Needed, The Enclosed Form May be Copied As Many Times As Needed.

PF-5

Certificate of Qualification

Notice: The following certificate of qualification is made pursuant to state law and _____ County ordinances. Vendor is cautioned to Carefully read this certificate prior to execution of this contract. Execution of the contract shall constitute execution of this certificate of qualification and shall also constitute a warranty by vendor that all the statements set forth within this certificate are true and correct statements of the vendor. Vendor is hereby notified that failure to execute this certificate shall result in disqualification from eligibility for the award of this contract. Vendor is further notified that in the event the county learns that any of the following certifications were falsely made, the contract shall be subject to termination.

A. _____ County Ordinance Chapter 10, Section 10.7.

Chapter 10, Section 10.7 of the Ordinance and Resolutions of the County _____ provides as follows:

10.7(1)—Persons and entities subject to disqualification. No person or business entity shall be awarded a contract or subcontract, for a period of five (5) years from the date of conviction or entry of a plea of *nolo contendere* or admission of guilt if that person or business entity:

(a) has been convicted of an act committed, within the State of bribery of attempting to bribe an officer of employee of a unit of state or local government or school district in the State in that officer's or employee's official capacity

(b) has been convicted of an act committed within the State of bid rigging or attempting to rig bids as defined in the Sherman Anti-Trust Act and the Clayton Act. 15 U.S.C. *et seq.*

PF-6

Appendix A—RFQ Boiler Plate

(c) has been convicted of bid-rigging or attempting to rig bids under the laws of the State;

(d) has been convicted of an act, committed within the State of price fixing or attempting to fix prices as defined by the Sherman Anti-Trust Act and the Clayton Act. 15 U.S.C. *et seq.*

(e) has been convicted of price fixing or attempting to fix prices under the laws of the State;

(f) has been convicted of defrauding or attempting to defraud any unit of state or local government or school district within the State;

(g) has made an admission of guilt of such conduct as set forth in subsections (a) through (f) above, which admission is a matter of record, whether or not such person or business entity was subject to prosecution for the offense or offenses admitted to;

(h) has entered a plea of *nolo contendere* to charges of bribery, price fixing, bid rigging or fraud, as set forth in subsections (a) through (f) above.

PF-6a

Certification of Qualification

Chapter 10, § 10.7(1) of the Ordinance and Resolutions of the County _____ provide as follows: § 10.7(1)—Persons and entities subject to disqualification. No person or business entity shall be awarded a contract or sub-contract, for a period of five (5) years from the date of conviction or entry of a plea of *nolo contendere* or admission of guilt, if that person or business entity:

(a) has been convicted of an act committed, within the State of bribery or attempting to bribe an officer or employee of a unit of state or local government or school district in the State in that Officer's or employee's official capacity.

(b) has been convicted of an act committed, within the State of bid-rigging or attempting to rig bids as defined in the Sherman Anti-Trust Act and Clayton Act. 15 U.S.C. § *1 et.seq.*,

(c) has been convicted of bid-rigging or attempting to rig bids under the laws of the State.

(d) has been convicted of an act committed, within the State of price-fixing or attempting to fix prices as defined by the Sherman anti-Trust Act and Clayton Act. 15 U.S.C. § 1 *et. seq.*

(e) has been convicted of price-fixing or attempting to fix prices under the laws of the State.

(f) has been convicted of defrauding or attempting to defraud any unit of state or local government or school district within the State.

(g) has made an admission of guilt of such conduct as set forth in subsections (a) through (f) above which admission is a matter or record, whether or not such person or business entity was subject to prosecution for the offense or offenses admitted to.

PF-6b

Appendix A—RFQ Boiler Plate

(h) has entered a plea of *nolo contendere* to charges of bribery, price-fixing, bid-rigging, or fraud, as set forth in subparagraphs (a) through (f) above.

I, _____ ,the _____ of Pres., sec., etc.

_____ having been duly sworn to
   Bidder/contractor/vendor

state the truth, do hereby swear the following to be true to the best of my knowledge:

1)_____ _____
   bidder/contractor/vendor   has/has not

been convicted, or entered a plea of *nolo contendere*, or made an admission of guilt to any act described in Chapter 10, § 10.7(1) (a) through (h) of the Ordinance and Resolutions of the County.

2) The owner, partner, or shareholder who controls, directly or indirectly, Twenty Percent (20%) or more of the business entity _____ been convicted or entered a plea of *nolo*
   has/has not

contendere or made an admission of guilt to any act described in § 10.7(1) (a) through (h).

3) _____ _____
   bidder/contractor/vendor   does/does not

employ as an officer, any individual who was an officer of another business entity at the time the latter business entity committed a disqualifying act described in § 10.7(1) (a) through (h); and

4)_____ _____
   bidder/contractor/vendor   does/does not

have an owner who controls, directly or indirectly, Twenty Percent (20%) or more of the business who was an owner who, directly or indirectly, controlled Twenty Percent (20%) of another business entity at the time the latter committed a disqualifying act described in § 10.7(1) (a) through (h). If any answer above is in the affirmative, I swear to the best of my knowledge that on _____ a court of competent jurisdiction
   date

PF-6c

entered judgement on the conviction of _____
                                      previous business entity
and/or the _____ that employed _____
                bidder/owner/officer
who is an _____ of the bidder/contractor/vendor.
              officer/owner

                                                                         _____
                                                                   President if Corporation
Subscribed and sworn to   <u>Owner</u> if Sole Propriety
before me this_____
day of _____, 19___   <u>Partners</u> if Partnership
                                                                 (or duly authorized
                                                                  Partner providing
                                                                   authorization)

_____
NOTARY PUBLIC

PF-6d

Special Provision

In Accordance With Public Act 85-1295, [as amended by Public Act 86-150] Section 33E-11 (Revised Statutes, Ch. 38, par. 33E-11).

The undersigned certifies that it is not barred from award of this contract as a result of a conviction for the violation of State laws prohibiting big-rigging or bid-rotating.

_____
Company Name

_____
Address

_____
Signature and Title

Subscribed and sworn to
before me this_____
day of _____, 19____

_____
NOTARY PUBLIC

PF-6e

The Undersigned Hereby Certifies That:

(1) the entity on whose behalf this certification is submitted has not been convicted, or entered a plea of nolo contendere, or made an admission of guilt to any act described in Chapter 10, Section 10.7(1) (a) through (h) of the Ordinance and Resolutions of the County _____.

(2) the owner, partner or shareholder who controls, directly or indirectly, twenty percent (20%) or more of the business entity has not been convicted or entered a plea of nolo contendere or made an admission of guilt to any act described in Chapter 10, Section 10.7(1) (a) through (h).

(3) it does not employ as an officer, any individual who was an officer of another business entity at the time the latter business entity committed a disqualifying act described in Chapter 10, Section 10.7(1) (a) through (h).

(4) it does not have an owner who controls, directly or indirectly, twenty percent (20%) or more of the business who was an owner who, directly or indirectly, controlled twenty percent (20%) or more of the business entity at the time the latter committed a disqualifying act described in Chapter 10, Section 10.7(1) (a) through (h).

B. Special Provision.

In accordance with Public Act 85-1295 [as amended by Public Act 86-150] Section 33E-11 (Revised Statutes, Chapter 38, par. 33E-11)

The Undersigned Hereby Certifies That: it is not barred from award of this Contract as a result of a conviction for the violation of State laws prohibiting bid-rigging or bid-rotating.

PF-7

Appendix A—RFQ Boiler Plate

C. **County Ordinance Chapter 10, Section 10-6.1.**

County Ordinance Chapter 10, section 10-6.1 provides that no person or business entity shall be awarded a contract or subcontract for goods or services with the County that is delinquent in the payment of any tax (including real estate tax) or fee administered by the County _____.

> **The Undersigned Hereby Certifies That:** it is not delinquent in the payment of any tax or fee administered by the County of _____ (including real estate tax) unless such tax is being contested in accordance with the procedures established by County Ordinance.

PF-7a

Execution By A Sole Proprietor

The undersigned acknowledges receipt of a full set of Contract documents and Addenda Number(s) _____ (None unless indicated here). The undersigned makes the foregoing Proposal subject to all of the terms and conditions of the Contract Documents. The undersigned certifies that all of the foregoing statements of the "Certificate of Qualification" are true and correct. The undersigned warrants that all of the facts and information submitted by the undersigned in connection with this Proposal are true and correct. Upon award and execution of this Contract by the _____ County Board of Commissioners, the undersigned agrees that execution of this Proposal shall stand as the undersigned's execution of this Contract.

Business Name: _____

Business Address: _____

Business Telephone: _____ FEIN/SSN: _____

    County Business Registration Number _____

SOLE PROPRIETOR'S SIGNATURE: _____

Date: _____

Subscribed and Sworn to

before me this _____ day

of _____, 19____.   My commission expires:

_____   _____
    Notary Public

      (SEAL)

**If you are operating under an assumed name, provide the _____ County Registration Number hereunder as provided by _____ Revised Statutes, Chapter 96, Section 4 *et sub.*

PF-8

Appendix A—RFQ Boiler Plate 197

Execution By A Partnership (And/Or A Joint Venture)

The undersigned acknowledges receipt of a full set of Contract documents and Addenda Number(s) _____ (None unless indicated here). The undersigned makes the foregoing Proposal subject to all of the terms and conditions of the Contract Documents. The undersigned certifies that all of the foregoing statements of the "Certificate of Qualification" are true and correct. The undersigned warrants that all of the facts and information submitted by the undersigned in connection with this Proposal are true and correct. Upon award and execution of this Contract by the _____ County Board of Commissioners, the undersigned agrees that execution of this Proposal shall stand as the undersigned's execution of this Contract.

Business Name: _____
Business Address: _____
Business Telephone _____ FEIN/SSN: _____
County Business Registration Number _____
Contact Person: _____

Signature of Partner Authorized To Execute Contracts on Behalf of Partnership:

*BY: _____ Date: _____

*Attach hereto a partnership resolution or other document authorizing execution of this Proposal on behalf of the Partnership.

Subscribed and Sworn to before me

this ___ day of _____, 19___.   My commission expires:

_____   _____
       Notary Public

(SEAL)

**If you are operating under an assumed name, provide the _____ County Registration Number hereunder as provided by Illinois Revised Statutes, chapter 96, Section 4 *et Sub.*

PF-9

Execution By A Corporation

The undersigned acknowledges receipt of a full set of Contract documents and Addenda Number(s) _____ (None unless indicated here). The undersigned makes the foregoing Proposal subject to all of the terms and conditions of the Contract Documents. The undersigned certifies that all of the foregoing statements of the "Certificate of Qualification" are true and correct. The undersigned warrants that all of the facts and information submitted by the undersigned in connection with this Proposal are true and correct. Upon award and execution of this Contract by the _____ County Board of Commissioners, the undersigned agrees that execution of this Proposal shall stand as the undersigned's execution of this Contract.

Corporate Name: _____
Corporate Address: _____
Contact Person: _____ Telephone: _____
FEIN: _____ Corporate File Number_____
*If the corporation is not registered in the State of _____ a current certificate of good standing is required from the State where your corporation is registered.

List All Corporate Officers:
President: _____   Vice President _____
Secretary: _____   Treasurer: _____
**Signature of President: _____
Attest: _____ (Corporate Secretary)
Subscribed and Sworn to before me
this ____ day of _____, 19__.   My commission expires:

_____   _____
     Notary Public

         (SEAL)

**In the event that this Proposal is executed by someone other than the President and Secretary, attach hereto a certified copy of the corporate by-laws or other authorization by the corporation which authorizes such persons to execute this Proposal on behalf of the corporation.

PF-10

Appendix A—RFQ Boiler Plate

Contract No._____
Proposal by a Corporation
The undersigned, hereby to be executed by a Corporation, acknowledges receipt of a full set of Contract Document(s) and Addenda Number (s) (None unless indicated here)_____.
  Addenda Number(s)

Corporate Name:_____
FEIN:_____
Corporate File Number:_____
By:_____
      (President's Signature)
Business
Business Address: _____ City _____ State ___ Zip_____.

(In the event that this bid is executed by other than the President, attach hereto a certified copy of corporate By-Laws or other authorization by the corporation which permits that person to execute this offer for the corporation).
      CORPORATE SEAL

Name of Bid Contract Person   List Officers
Contract Person:_____   President_____
           (Print Name)
Telephone No.: _____   Vice President_____
           (Print Name)
         Secretary_____
All Officers of the Corpora-   (Print Name)
tion must be recorded-If not   Treasurer _____
applicable please indicate   (Print Name)
         Attest:_____
         Secretary's Signature

— — — — — — — — — — — — — — — — — — — —
Subscribed and Sworn To Before Me This
_____Day of_____, 19___.
            Notary Seal
_____   Commission Expires:_____

PF-10a

CONTRACT NO._____

Proposal Acceptance

The undersigned, on behalf of _____ County, a body politic and corporate of the State hereby accept the foregoing bid items as identified in the proposal.

Total Amount of Contract: $ _____
(Dollars and Cents)

Fund Chargeable: _____ Department: _____
(Fund Number)   (Number)

Dated at _____ this _____ Day

of _____, 19 _____

_____
President, County Board of Commissioners

_____
County Comptroller

_____
County Purchasing Agent

Approved as to Form:

_____
Assistant State's Attorney

PF-10b

Appendix B

Preventive Maintenance Assignments

(These assignments are purposefully general in scope to cover a wide range of applications and equipment types. They are intended to provide a data base of information from which Maintenance Managers can draw to get quickly up and running. Once installed, the procedures can be added to or deleted from PM work orders to suit his/her particular operation and frequencies as established.)

Air Compressor
Inspect for obvious defects
Run and observe for proper function of indicators.
Calibrate gauges
Clean unit
Clean or change filters
Check motor and pulley alignment
Record pressure drop across main air line filters
Inspect for oil leaks
Check and record oil level
Check V-belts adjusting tension as needed
Record amps and voltage with motor operating
Record compressor cut-in and cut-out pressure
Clean condenser coil and fan motor on air dryer
Oil condenser fan motor on air dryer
Inspect unit for leaks
Check pressure regulating valve operation
Drain moisture from air receivers and traps

Manually lift safety relief valves
Check operation of unloaders
Check all electrical connections for tightness
Inspect piping for leaks
Clean air intake filter
Test automatic safety devices
Drain water from volume tank and traps
Operate unit to insure proper function
Check shaft seals and packing

Air Handling Unit
Check vane and damper operation
Check low limit alarms
Check belts for proper tension and excessive wear
Insure all guards are in place
Check for proper operation
Check pulleys for tightness and excessive wear
Check mounting bolts and fan wheel setscrews for tightness
Record amperages and voltage with motor operating
Inspect and clean chilled water coils, condensate pan and drain
Check humidifier operation and dispersion tub
Inspect and tighten all electrical connections
Check contacts and contractor operation
Clean entire unit and plenums
Inspect filters and replace as needed
Check fan and housing for loose bolts and setscrews
Check fan for tightness and remove dirt accumulations
Lubricate bearing and fan motor
Check v-belt drives and belts for alignment and wear
Check damper operation and lubricate as required
Check bearing, setscrews and bolts and note vibration
Clean fan impeller
Wash preheat and cooling coils
Calibrate operating controls
Calibrate safety controls
Calibrate temperature controls

Boiler
Secure boiler for annual inspection
Shut the main steam valves and feedwater valves

Appendix B—Preventive Maintenance Assignments

Clean boiler water sides
Clean boiler fire sides
Inspect water column and gauge glass
Test feedwater makeup control and pump operation
Inspect feedwater system
Operate safety devices and check control operation
Test low water fuel cut-out
Clean entire unit, piping and air intake louvers
Manually operate safety valves
Test automatic safety devices
Check burner linkage
Check pump couplers
Check packing valves and pumps
Check fire eye scanner
Inspect seals on boiler access doors for air leakage
Inspect boiler tubes and tube sheets
Inspect boiler enclosure refractory and insulation for damage
Inspect and lube boiler access door hinges and latches
Repair leaking steam fittings, valves and feedwater components
Replace valve packing as necessary
Inspect all water, steam and fuel lines
Check soot blowers for proper operation
Change out all gaskets
Replace fusible plug
Check baffles, firebox and firesides
Inspect brick work and burner cone
Reassemble unit and clean exterior
Lubricate combustion blower motor and all damper linkages
Set fuel/air ratio for efficient combustion
Record CO_2 and O_2 readings from flue gas analysis test
Dismantle and inspect all appliances
Calibrate burner programmer
Inspect breeching and stack
Check chemical pump system
Inspect condensate return system

Chiller
Check oil level; remove and clean strainers
Check refrigerant sight glass for flow
Inspect for obvious defects and note

Inspect timers, relays, switches and wiring for defects
Check contacts
Clean air cooled condensers
Inspect piping for leaks
Operate unit to insure proper function
Check and clean coils
Check and replace filter and belts as needed
Lubricate fan motors
Check and note pressure readings
Adjust purge drive belt tension
Check magnetic starters and clean starter panel
Check all electrical connections for tightness
Check safety controls and log cut-in and cut-out points, low oil pressure, condenser, high-pressure cut-out, chilled water low temperature cut-out, anti-cycle time delay and flow switches
Check guide vane operation
Check demand limiter and transformer
Check compressor windings with megohm meter
Check oil pump Check lube oil sump heater
Calibrate operating controls
Calibrate all safety controls
Tighten all loose connections
Inspect bearings and heat transfer surfaces
Test relief valves
Troubleshoot control circuits

Cooling Tower
Visually inspect entire unit for deterioration
Inspect motor pump coupling for alignment and tightness
Lubricate motor and pump
Check gearmotor assembly and gear reducer for oil leaks
Change oil in gear reducer
Check V-belt tension and pulley alignment
Check spray tree nozzles and drift eliminators
Brush tower slats to remove solids build-up
Clean debris from inside tower
Check floats
Ream out all orifices
Adjust bleed valve rate
Check operation of make up valve

Appendix B—Preventive Maintenance Assignments

Observe fan operation
Clean basin and sump
Check fan and motor mounts for tightness
Remove debris from air louvres
Check for belt slippage
Check make-up float valve operation
Inspect all motor electrical controls
Operate unit and check for noise/vibration
Record amperage and voltage of motors
Check the overflow to see that it is free
Examine valves and valve seats for wear
Check alignment of the fan drive shaft
Check gaskets for damage and replace
Replace defective screens and strainers
Check the general condition of structural members
Tighten loose bolts and connections
Check pipes and metal parts for corrosion and rust
Inspect for leaks
Check oil level in gear boxes

Electric Motor
Check for loose terminals and faulty wire insulation
Check for signs of overheating
Check ground connections
Check switches and breakers for proper mounting
Check contacts
Check fuse sizes
Check thermal and overload relays
Clean panel interior
Observe motor in operation, checking for noise and vibration
Check mounting bolts for tightness, wear and misalignment
Check for evidence of overheating of motor or bearings
Record amperage and voltage
Lubricate as prescribed by the manufacturer
Inspect for cleanliness and blow out with moisture free air
Observe operation for chatter or hum and evidence of bad coil or poor contact
Operate start/stop button for free movement
Check conductors for overheating
Clean foreign accumulations on windings and air passages

Check all electrical connections for tightness
Check the condition of coil insulation and examine all windings
Check bearing wear and rotor clearances
Clean out and renew grease in all ball and roller bearing housings
Measure insulation resistance by the dielectric absorption or high potential test method
Open-frame motors in dusty or linty locations should be cleaned with vacuum equipment unless designed for cleaning with low pressure compressed air
Check oil level sleeve bearings and condition of oil rings
Check the bearing temperature
Inspect motor surroundings for water, oil, steam, dirt, dust and any loose objects
Motors with commutators or slip rings should be checked for excessive sparking. Examine brushes for excessive wear and condition of holders. Collector rings should be clean and smooth with no evidence of scoring or pitting

Electric Panel
Check panel for overheating
Check for loose terminals and faulty wire insulation
Check wiring for signs of overheating and for ground integrity
Check switches and breakers for proper mounting
Check panel doors, hinges and latches for proper fit and operation
Replace all panel screws
Inspect and clean dust or lint with dry compressed air
Remove any foreign materials
Clean corrosion from fuse contacts
Cover all unused conduit openings
Check condition of doors, covers, gaskets, etc.
Check necessary grounds
Check ventilation where necessary
Replace oversized fuses with those having the proper ampacity
Check the trip setting of all circuit breakers
Be sure that panels are not exposed to rain, snow or other liquids
Operate each switch a few times by hand with no load
Examine all devices for missing or broken parts, excessive wear, and proper spring tension
Clean all copper contact surfaces
Lubricate the operating parts of all switch mechanisms

Tighten all bolted bus connections
Check for loose electrical connections and mechanical fastening
Check for accumulations of dust and dirt
Check for structural distortion, overheating and poor or loose connections in the bus bars
Check for obstructions to movement of mechanical parts
Check for control wiring condition

Elevator
Operate car and make general check of car interior
Check emergency telephone operation
Verify that permit to operate is current
Inspect equipment room
Note any corrective maintenance needed
Operate car and counterweight safety-hand pull out
Check governor operated car and counterweight safeties
Check governor operated car safety-rated-load test at rated speed
Follow instruction in manufacturer's manual for raising and servicing
Wipe off hydraulic rams and pistons.
Apply small amount of oil to pistons
Lubricate at grease zerks and hinge pins
Check hoses and connections for fluid leaks
Check pit and remove trash and debris.
Check for proper drainage
Check machine for "soft start" and smooth acceleration
Check mechanical and electrical alignment and adjustment of each station
Check motor brake Inspect motor starter
Check drive motor for excessive noise and/or vibration.
Check motor and gear box for overheating
Check drive and car-carrying chains for cleanliness and proper lubrication
Check station selector lights at each floor
Dispatch and receive a container at each station.
Check that operation is smooth and free from vibration and excessive noise.
Observe the operation of all limit switches

Fan
Inspect unit visually for defects
Check for excessive vibration or noise

Lubricate fan grease fitting with a hand operated grease gun
Check belts for wear, cracking, etc. (Belt tension should be adjusted so that when belt is depressed in center it should depress its width)
Rotate motor and fan manually
Check all motor guards, pulleys, thrust collars and bearings
Check for excessive wear
Clean motor, fan blade and guards as needed
Inspect all electrical connections for tightness
Check and clean magnetic starters
Lubricate motor and fan bearings
Clean entire unit
Check connecting ductwork for air leaks
Record amperages and voltage with motor operating
Check damper operation
Check pitch of fan blades
Conduct lubricating oil trend analysis
Inspect for abrasive erosion, corrosion and deposit buildup
Grease anti-friction bearings
Change oil if operating under adverse conditions
Check axial, horizontal and vertical vibration
Check coupling, shaft alignment and clearance between stationary and moving parts
Check foundation bolts for tightness
Clean motor, pulleys and thrust collars
Replace all guards and panels

Fire Warning System
Check for loose terminals and faulty wire insulation
Check for signs of overheating
Check ground connections
Clean panel interior
Check indicator lamps and replace as needed
Test fire alarm operation
Check all standpipe connections for evidence of leaking
Check the packing on all valves for leaking
Observe heads for any abnormal conditions
Observe heads for signs of leakage
Check and insure that normal flow pattern has not been obstructed
Check all fire hose connectors and valves for leakage

Ensure pressure gauges are functional
Check for paint chipping, cracked glass, etc.
Observe and note suction and discharge pressures of fire pumps
Assure coupling guard is in place and fasteners are tight
Check pump suction and discharge valves
Check panel indicator light to assure electric power to pump
Check pump packing glands
Check that control valve is secured in the open position
Inspect all other control valves
Check valves for proper seal open position and leaks
Check gauges for proper pressure
Test alarms by opening the inspector's test connection and perform a water-flow test
Secure sprinkler system control valve in its normal or open position
Record condition of valves—open or closed, properly sealed, in good operative condition, turn easily, do not leak, readily accessible
Check each inspector flow pipe to see if water is at the proper level in sprinkler system
Ascertain and record all pressures checked
Check operation of all detectors

Generator
Check vibration of machine before and after the overhaul
Test the insulation resistance of the windings
Check and replace plugs, points, condensor, rotor and plug wires
Clean all relays with liquid burnisher
Clean generator exciter
Check brushes for excessive wear and clean same
Change oil and filters
Megger exciter field and armature
Check all oil lines for leakage and worn surfaces
Check coolant level and fill to normal level
Check batteries and charger
Vacuum main and dc breakers
Clean voltage regulators
Clean motor control panel
Remove dust from collector ring insulation and brush rigging
Check brushes for freedom of movement, excessive wear, pressure and alignment

Check collector ring for smoothness of operation
Inspect the lube oil system for possible leaks, excessive vibration and over temperature
Check the air gap on sleeve bearing motors
Check sleeve bearings clearances and inspect for wear
Check the end play of motor and generator shafts
Inspect brushes for wear and heat cracks, checking for fit and free play in the holders
Check brush spring pressure
Examine commutator for high bars, high mica, wear and burning
Thoroughly inspect all ball, roller and sleeve bearings for wear and dirt
Check all connections and terminals for tightness and corrosion
Check continuity of ground

Heat Exchanger
Visually inspect for defects and leakage
Check control setpoint and control valve operation
Establish design flow and level in and through heat exchanger
Check unit performance Inspect tube sheet and tubes for corrosion and scale
Chemically or mechanically clean as needed
Replace gaskets as needed
Check heating element, safety valve, solenoid valve, steam trap, pressure control switch and drain
Clean all strainers and flush dirt legs
Operate and observe for proper function of indicators
Calibrate temperature, operating and safety controls
Inspect all piping and fittings
Check float operation
Service auxiliary devices Inspect all surfaces for deterioration
Repair and/or replace broken stays and supports
Record inlet and outlet temperatures
Record inlet and outlet pressures
Check unit for unusual noise and vibration
Check tightness of heads, bolts... etc.
Replace leaking gaskets
Ascertain vents are open
Check operation of valves and floats
Check cleanliness of strainers

Incinerator

Remove all ashes from incinerator interior
Inspect interior of chamber for loose or missing refractory material
Operate safety devices
Check control operation
Check flame safeguard unit for positive fuel valve shutoff
Check firebox, baffles and brickwork
Check stack mounts, stack supports, and ash screen
Lubricate gearings on combustion air blower motor
Clean burner and gas passages
Check combustion blower, motor, damper operation, combustion control valves, linkages and regulators. (Lubricate as needed)
Operate and observe for proper function of indicators
Check flame and burner air fuel adjustment
Check fuel valve operation
Inspect and lubricate loader hopper door assembly
Inspect conveyor drive gear box and mounts
Check drive chain for proper tension
Check operation of water level devices
Lubricate all ash conveyor bearings
Lubricate modulating motor linkage
Inspect and maintain ash conveyor system
Inspect spray nozzles
Inspect and maintain incinerator upper chamber
Inspect and maintain incinerator limefeed system
Clean out lower chamber of accumulated residue
Remove excess slag from lower chamber
Inspect lower and upper chambers for damage
Inspect hearth for excessive wear
Inspect front of transfer rams for wear and leaks
Inspect underfire air tubes for wear and leaks (Clean all tubes)
Inspect and verify operation of water spray nozzles
Inspect and clean thermocouples
Inspect tap-in for excessive wear or warping
Inspect all access door gaskets (replace segments as necessary)
Perform incinerator burner maintenance
Run fan check for vibration
Inspect and lube front and rear bearings
Check oil level in gear boxes
Inspect and maintain bag house and air lock

Miscellaneous Electrical
Make a complete inspection of all wiring
Check ampere load on all circuits
Check and record insulation resistance
Check windings for dirt, moisture, cracks and loose wedges
Check coupling alignment
Clean and paint corroded conduit
Support any loose wires
Protect wires subject to physical damage
Check necessary ground connections
Check wiring for accessibility
Check for dirt, grease, carbon or metal dust accumulations
Check for burrs, or fused metal globules on contacts
Check for loose connections
Check for sluggish or obstructed movement of moving parts
Check contacts and replace if necessary
Inspect bearings for wear
Check calibration of and operate relays
Assure adequate ventilation is provided for all battery storage areas to prevent hydrogen accumulations
Inspect battery terminals to make sure that they are clean, tight and free of corrosion
Dress rough spots on contacts with sandpaper (blow off grit)
See that latches and triggers are properly adjusted
Inspect copper arcing tips (Dress as necessary).
Clean and tighten all connections and lubricate bearings
Check contact alignment and adjustment
See that latches and triggers are properly adjusted
Examine main current paths for evidence of overheating
Check pins, bolts, nuts and general hardware
Check control wiring for loose connections
Check settings for auto tripping units
Check reliability and adequacy of circuit breaker

Miscellaneous Mechanical
Visually inspect for obvious defects and note findings
Clean and inspect electric motors
Clean fan impeller
Lubricate fan and motor bearings
Check for vibration

Appendix B—Preventive Maintenance Assignments

Clean oil strainers
Test automatic safety devices
Clean unit exterior
Change lube oil and filters
Inspect V belts
Operate unit to insure proper function
Check fluid level in the batteries
Clean heat exchange surfaces
Tighten all connections
Check alignments
Reinstall access panels and covers
Check door latches
Inspect timers, relays and switches
Inspect valves for proper operation
Check piping and vessels for leaks
Calibrate pressure and temperature controls
Ascertain reliability of safety devices
Check foundation and mount integrity
Lubricate moving parts as required
Check bearings
Check fluid levels and overflow piping
Change air filters
Check wiring integrity
Clean and or repair trap system

Package A/C Unit
Check electrical connections, fan motor bearings and mounts
Check compressor and crankcase heater operation
Calibrate pressure gauges and thermometers
Check operation of all time delays, relays and contactors
Check high and low pressure cut outs
Clean and lubricate condenser and evaporator fans and motors
Check condensate tray for proper drainage
Clean unit interior and exterior
Reinstall access panel screws
Change or clean filters as needed
Check dampers, V-belt and condenser and evaporator coils
Record amperages and voltage with compressor operating
Inspect On/Off and limit switches
Check high pressure and low pressure cutouts

Clean condenser and evaporator fans and motors.
Lubricate fans and motors as necessary
Check condensate tray for proper drainage
Clean condenser and evaporator coils
Check electrical connections for tightness
Check all fan motor bearings and mounts
Check integrity of electrical wiring
Check compressor operation, mounts, and oil level on semi-hermetics
Check control thermostat operation
Inspect piping for leaks
Check crankcase heater operation
Check refrigerant charge-note compressor suction and discharge pressure in relation to supply, return, and outside air temperatures
Check refrigerant levels in sight glass
Megger compressor windings and note findings

Pumps
Check for unusual noise and vibration
Ascertain pressures and temperatures are correct
Check for abnormal flow conditions
Check motor, pump alignment and flexible coupling
Check suction and discharge gauges
Check shaft seal packing for leaks
Clean motor, pump, piping and surrounding area'
Check bearing temperature
Check packing for sufficient leakage
Inspect plunger for scoring
Inspect gaskets and diaphragms
Check for obvious wear and damage
Check wearing clearances
Check condition of foot valves and check valves
Inspect all electrical connections for tightness
Clean and check magnetic starter and temperature controls
Lubricate unit
Clean and flush suction strainers
Record amperages and voltage with unit operating
Check operation of pump discharge check valve
Check oil level
Check unit for unusual vibration
Check V-belt (adjust or replace as needed)

Check pump rotation for direction and free operation
Calibrate pressure gauges, thermometers and flowmeters
Check impellers for corrosion, erosion or excessive wear
Perform recommended checks on driving motor
Inspect motor pump coupling for alignment and tightness
Manually operate relief valve
Inspect piping for leaks

Refrigeration
Clean unit exterior
Check electrical connections
Clean water strainer
Check piping insulation and capillary tubes
Remove dirt and lint from condensers
Grease bearings
Clean and adjust float valve
Check fan motor bearings
Clean compressor
Clean evaporator coil
Record amperages and voltage of fan motors
Check discharge and suction pressures
Check electrical connections
Check liquid line solenoid valve operation
Inspect piping insulation
Check refrigerant sight glass for proper charge
Check compressor oil level with compressor running
Inspect door gaskets
Inspect piping for leaks
Inspect On/Off and limit switches, wiring and connections
Check power to unit
Inspect and clean refrigeration components
Inspect door hinges and latches
Check defrost cycles
Disinfect areas contacted by food and water
Measure ground wire resistance
Check operation of temperature alarm
Check box lights and door switches
Record temperature drop across coils
Check electrical connections for tightness
Add charge to unit as required

APPENDIX C
SPECIAL OPERATING CONCERNS

There are four major areas, outside of those we've already discussed, that will be of primary concern to you in the operation of your plant. They deal with the combustion process, water analysis and treatment, lubrication requirements and corrosion. Entire volumes have been written on these topics and it would behoove you to include them in your engineer's library for later scrutiny. For now, it will suffice that we address the relevance of the issues. What do you say we jump into water treatment to get us started?

WATER TREATMENT

The Natural Water Cycle

Anyone who's been caught outside during a cloudburst has a good idea where water comes from. The oceans evaporate, creating clouds of water vapor which ultimately give up their moisture in the form of rain or snow. A quick and easy explanation, you say; but that's only part of the story. Up until now I was referring to relatively untainted water in its various states. The truth is water in its natural state is never pure. From the time it begins to precipitate back to earth, it wages a losing battle against contamination. As it's hurled earthward it picks up airborne particles and gasses. After touching down, it dissolves and absorbs a multitude of constituents as it erodes across the face of the planet on its trek to the sea. And not all of it completes the journey. Some of the water ends up in bogs, lakes, or ponds, some of it replenishes the ground water we tap into with our wells, and much of it is consumed by animals or vegetable life or evaporates from drying riverbeds and our own backyards.

Constituents of Water

One thing is certain: by the time it gets to you, the water you use in your equipment is a veritable slurry of organic and inorganic compounds and dissolved gasses. Depending on where it is found, it may be teeming with life or completely devoid of it, as the result of chemicals added to it by man. As the plant engineer you are responsible for determining what impurities are contained in your water, ascertaining what effect they can have on your operation and implementing a water treatment program for their control or removal.

Associated Problems

The most prominent problem associated with the use of water in equipment is corrosion—in the form of metal deterioration, scale formation—resulting in poor heat transfer and metal fatigue and fouling, a condition whereby passages, pipes and nozzles become clogged due to the accumulation of the impurities deposited within them. No less important, poor-quality water also results in foaming, whereby contaminated water forms a froth resembling soap suds in the steam space of boilers. Priming, a consequence of foaming, is the carryover of impurities and water into steam discharge lines. And embrittlement (a condition of metal evidenced by hairline cracks which develop in high stress areas exposed to alkaline salts).

Removing Impurities

On the basis that the water you use at your plant is purchased from your local sewage treatment facility, it's likely that most of the suspended matter originally contained within it has been filtered out using some combination of clarifying process including sedimentation, coagulation and/or flocculation. How's that? You're right, it's getting too involved and technical-sounding, but I'm on a roll, so don't bother me. My intent here is point out that regardless if the water you purchase from your local authority is aerated, chlorinated, fluoridated or carbonated, it's quality still doesn't meet the strict operating standards of your equipment. Though most of the suspended solids have been removed, the dissolved salts and gasses remain, waiting to be manifested as scale or corrosion. The three major tools used in the physical plant for improving water quality are demineralization (the removal of inorganic salts from solution by ion exchange), deaeration (a process in which dissolved oxygen and carbon dioxide are removed by heating of the water), and internal treatment (a process whereby a variety of chemicals and techniques are used to condition the water to pre-determined values).

Program Benefits

Most organizations, large enough to employ viable operating engineers, take advantage of their professional knowledge to help defray operating costs in the physical plant. Such is the case with water conditioning. Analysis and treatment is an area that can pay handsome dividends if properly performed. The by-products of a well implemented program are increased heat transfer efficiencies, lower fuel expenditures and decreased consumption of chemicals. Except for steam trap maintenance, it has the most potential for reducing annual operating costs in the power plant. If you don't have the expertise in-house, by all means contract it out to a reputable vendor.

Service Agreements

Depending on the magnitude of a company's operation, its policy on contractual arrangements and often times, the political environment, you may or may not be regularly serviced by the people from whom you buy your water treatment chemicals since many companies' purchases consider only the bottom line. The fact is the decision from whom to buy shouldn't be based on cost alone, but on the service you receive from the vendor. Quote me! "Chemicals are chemicals," regardless if they are referred to by their generic nomenclature, their brand name or an alphanumeric designation. Quote me, again! "Any company that provides service is only as reputable as their representative who services your account." Once you find a good one, don't let him get away. The extra money you spend to get and/or keep him will be returned many fold.

Corrosion

Without a doubt, corrosion is one of the biggest headaches facing the operating engineer and the older his plant, the more dedication he'll need for its treatment. I say treatment instead of cure because man hasn't found a way to totally eliminate the problem, only ways of minimizing its effects. In the physical plant, corrosion is the common ground where Father Time, Mother Nature and Mr. Murphy most like to meet as a group. Over time, the metals used to construct your vessels, equipment parts and system components slowly rust in an attempt to revert back to the natural ores from which they were manufactured. The process begins when metal ingots are formed into billets and blooms which are turned into the technological marvels we operate in our spaces. It continues as the carcasses of our aged devices are cast aside onto scrap heaps and only ends when the once-magnificent machines have returned to dust. And while Nature and Time get the corrosion process underway, Murphy just waits in the wings figuring out ways to make your life the most miserable, at the worst possible moments.

The Process

Corrosion is basically an electrochemical process in which electricity flows through a solution of ions between areas of metal. Deterioration occurs when current leaves the negatively charged metal (anode) and travels through the solution to the positively charged metal (cathode) completing an electric circuit, much like the action in the cell of a battery. The anode and cathode can be two different metals or different areas of the same piece of metal. Corrosion occurs when a difference in electrical potential exists between them or as the result of their physical contact. The solution containing the ions is called the electrolyte, which in most cases is essentially comprised of water, its ions and the ions generated by the anode.

Effects of the Process

Although the process remains constant, corrosion is manifested in different ways. The way a particular piece of metal is acted upon depends on its material composition, the temperature and pressures to which it is subjected, the environment in which it is contained and a myriad of other variables, too numerous for consideration in this light dusting of the topic. Generally speaking, types of corrosion are evidenced by the effects of the corrosion process under varying conditions. Pitting, for example, is a localized form of corrosion in which small pockets or voids develop in the surface of metal as the result of the breakdown of protective films or from oxygen concentrations. Uniform corrosion is a general wasting of the metal over large areas of its surface caused by high temperature oxidation, exposure to acidic solutions or containment in corrosive atmospheres. And stress corrosion cracking results from static loading of metal surfaces having built up tensile stresses. Corrosion occurs where there are high levels of oxygen or carbon dioxide, low pH values, where contact is made between dissimilar metals and in damp environments or corrosive atmospheres.

Appendix C—Special Operating Concerns

Assessing Your Involvement

Don't wait until your plant takes on the appearance of a roadside auto graveyard before deciding you may have a corrosion problem. I submit that the problem began long before your company opened its doors for its first day of business. Corrosion is a constant and unrelenting process that continues until the metals manufactured by man have been returned to the ores from which they were produced. You should check out your equipment before it checks out on you. No metal structure or device should escape your scrutiny. Where should you begin? There is no best starting point; I suggest that you do it in three stages. In the first stage, visually observe the entire plant, detailing which areas show evidence of a corrosion problem. In stage two, open units to internal inspection and review their records of operation, lubrication and water treatment. Finally, send scrapings and samples of corroded parts out for laboratory analysis. The information gathered during the evaluation process can be compiled and studied and a plan of corrective action can then be formulated.

What To Look For

The bulk of your detective work will be accomplished in the second stage of your investigation, as it is the interior metal surfaces of your equipment that are most subject to the ravages of corrosion. The insides of our machines and vessels are frequently exposed to extremes of pressures and temperatures. Many are cooled by or use water in their operation. Their structures expand and contract as they heat up and cool down. And some are adversely stressed as the result of loads imparted on them.

Of the many causes of corrosion, chief among them are oxygen and carbon dioxide liberated from heated water resulting in grooving and pitting. This condition is most evident in the steam and condensate lines of boiler systems. The main culprits in refrigeration units are the calcium and sodium brines used as secondary refrigerants that attack the lines and devices through which the solutions circulate. Improper water treatment can result in poor corrosion control which in turn can cause choking off of narrow passages, nozzles and orifices.

Often times much can be told about the condition of a device by combining a review of its treatment record with an observation of its interior. For instance, if you were to open a fire-tube boiler and found it to be completely clean and devoid of scale, you might consider it to be in excellent condition. But if you were to review the water treatment log for the unit and find that the water contained within it had an average pH value of 6.0 over the past year, you might want to consider whether or not your boiler has been in a general condition of uniform corrosion (caused by people on the day shift, of course) and that the vessel may be suffering from a severe loss of metal, subsequently lowering its pressure integrity.

There are a number of critical things to look for when checking for corrosion in your plant, but as you can see from the last example, probably none more important than good, sound advice from an expert. Any time you open a piece of equipment in your plant, it's always a good idea to have an inspector or manufacturer's representative there to provide you with his opinion of any problem found.

Limiting Your Exposure

Obviously, the best way of defending your equipment's insides against the damage caused by corrosion is by properly filtering, treating and deaerating the liquids you inject into them. But what about the exterior of those units? You can limit your problems by shopping around for units manufactured from corrosion-resistant metals or substituting non-metallic-constructed units for traditional metal ones when purchasing new; or you can accept the fact that corrosion will always be with us and draw up a plan for managing it. Depending on how much you estimate your potential savings might be, you may or may not want to have a consultant hired to study your plant and design a program, but in any case you should make corrosion prevention and control a priority in your operation. Some things that you can do to lower the incidence and severity of corrosion are:

- lower the operating temperatures of moving parts
- substitute corrosion-resistant materials
- limit loading stresses on your equipment
- use only high-quality stable lubricants
- keep different metals from physically touching
- strictly adhere to all PM guidelines
- maintain water quality within recommended parameters
- keep all connections tightened and repair corroded areas
- filter out contaminants in equipment fluids
- don't over tighten flange and connection bolts
- eliminate sources of vibration

- maintain pH levels within set limits
- limit the conductivity potential of plant waters
- deaerate feedwater before introduction into your boilers
- maintain proper stack temperatures
- use cathodic inhibitors in your treatment program
- lower relative humidity in the plant
- redirect corrosive fumes away from equipment
- keep spare parts packed in grease until used
- lower the speed at which liquids are transported
- use protective primers on all metal surfaces
- repair leaks and replace bad insulation
- use neutralizing and filming amines in water systems
- keep non-metallic liners in metal tanks in good repair

COMBUSTION

The Mechanics of Burning

Have you ever seen someone throw a lighted cigarette into a bucket filled with gasoline? I have. No, it didn't explode; it didn't even catch fire. As a matter of fact the gasoline put the cigarette out. Please don't attempt this yourself because if it's done improperly the gasoline can catch fire, very easily! So why didn't a fire start when I witnessed the event? Because something was missing. There are two basic premises in combustion theory:

- combustion only takes place on a vapor level
- a specific ratio of fuel vapor and oxygen, at the proper ignition temperature must exist for combustion to occur

So what was missing? Perhaps there wasn't enough vapor being generated or the heat from the cigarette wasn't in contact with the vapor long enough. I'll never know for sure. You see, the experiment wasn't performed in a laboratory under controlled conditions by a qualified combustion expert. It was done at a local service station, near the fuel pumps, by a 16-year-old whose mental prowess was suspect. Not that he couldn't get away with extinguishing a cigarette in this manner time after time, because he did, but I contend that he was lucky... very lucky, to tempt fate and win as he did.

What if the outdoor temperature were higher that day and the volume of vapors given off by the fuel were substantially increased? What if the tip of the cigarette broke off and its ember was atomized as it passed through the volatile vapors? What if the wind shifted as the ember passed through the vapors. The vapor theory is the reason why candles can't burn without the aid of a mechanical wick; different size grates and combustion chambers are specified for high and low-volatile solid fuels, and why fire extinguishers are so effective at extinguishing large fires when their contents are discharged at the base of the flame.

Fuel In The Plant

Outside of manufacturing industries that use exorbitant quantities in their production processes, Stationary Engineers are probably the largest commercial consumers of fuel in North America. They burn every type imaginable and many have plants that are veritable fuel depots. Tanks containing propane and butane dot the landscape of facilities that don't have access to natural gas pipelines. Fuel oil tanks are located above or buried below ground for supplying boiler furnaces when their normal gas supply is interrupted. Diesel fuel is stored for use in emergency generators used to provide electricity when utility power is lost. Gasoline pumps supply fuel for operating plant vehicles.

Plants that use solid fuels, such as coal, wood and even refuse, stockpile their supply in bins, often maintaining reserves of several months. In addition to the autonomous fuel banks they may possess, most plants purchase fuel from their local utility companies. Natural gas is piped into the plant to fire boilers, domestic hot water tanks, space heaters and dietary food processing equipment. Electricity, though usually not considered a fuel itself, is used for many of the same purposes as well as for resistance heating of fuels, baseboard comfort heating and illumination.

Characteristics Of Fuels

Fuels come in three physical states—solid, liquid and gas. As the combustion process is the rapid oxidation or combining of fuel atoms with oxygen atoms, fuels must first be reduced to their vapor form before they can be burned. Gaseous fuels readily combine with oxygen and need only be brought up to ignition temperature. Properly oxygenated, they are generally clean burning, leaving little or no residue. Approximate heating values of gases range from 1000 to 3000 BTU's per cubic foot. Liquid fuels must be transformed to their gaseous state before being burned but once vaporized, generally share the same requirements as those for gaseous fuels.

The portion of the liquid fuel that is not vaporized is evidenced by the carbon it deposits throughout the combustion chamber and gas passes of the unit in which it is burned. The principal liquid fuel is oil which is referred to be grade as light, number 1 and 2; medium, number 3 and 4; or heavy, number 5 and 6. Approximate heating values range from 135,000 to 156,000 BTU's per gallon. Solid fuels must be heated to release the volatile materials contained within them. ' They are the most difficult fuels to prepare for the combustion process and leave behind large quantities of ash when burned.

The principal solid fuel is coal, but anything that burns can be used as a fuel if your equipment is designed to handle it. Approximate heating values of coal range from 7,000 to 15,000 BTU's per pound.

Combustion Theory

Up to this point we have referred to the combustion process as rapid oxidation of a fuel source. Whereas that is a true summary of what occurs during the fuel-burning process, it is also a misleading statement because pure oxygen itself is not flammable; it only supports combustion. And the purer the oxygen is that's combined with a fuel, the hotter and faster the fuel will burn. As the purity of oxygen and its ratio to fuel increases, it ceases to be a catalyst for combustion and becomes an ingredient of an explosive mixture.

So it's just as well that the oxygen we use for burning our fuels in the physical plant is derived from the same diluted air we breath which, by weight, contains about 21 percent oxygen. During combustion, the oxygen derived from the air combines with fuel vapors resulting in the liberation of thermal energy. To what extent fuel is consumed is dependent on several factors. If perfect combustion is your objective, chances are you'll be disappointed with the results.

Theoretically, PERFECT COMBUSTION can only occur when each atom of fuel is brought into contact with the exact amount of oxygen needed for its consumption at a precise temperature and held there long enough to complete the process. But if your objective is simply to complete the combustion process, your chances are better.
COMPLETE COMBUSTION is accomplished by supplying more than the theoretical amount of oxygen into the fuel mixture to assure full consumption of the fuel. In boiler furnaces, this is accomplished by supplying "primary" combustion air through air registers at the

burner, then introducing "secondary" air to aid in mixing the air with the fuel. This results in the total consumption of the fuel, but some oxygen is always evident in the exhaust gasses.

The higher volume of air supplied is referred to as "excess" air and is a function of the draft system INCOMPLETE COMBUSTION happens when one leg of the fire triangle is affected; either the fuel mixture is too "rich" or too "lean," it hasn't reached ignition temperature or it isn't sustained for a long enough period of time.

LUBRICATION

Whether it's carried out by the operating engineers, preventive maintenance mechanics or persons designated as oilers whose sole responsibility is to keep their machine's moving parts moving smoothly, a properly implemented equipment lubrication program is essential to the economical operation of a power plant. And here are some things he should know before he grabs his oil can and indiscriminately starts squirting.

Its Basic Function

So why are we always oiling our machines? The answer seems simple enough: if we don't, they'll gradually bind up and eventually stop running altogether. But the need for lubrication can't be that easily explained away because there's more to it than just making metal surfaces slippery. Aside from reducing friction, lubricants perform in a variety of other ways. They are used to flush out contaminants that accumulate in oil passages and reservoirs. When applied to metal they inhibit the corrosion process. Trapped in the mesh area of gear teeth they provide a cushioning or shock absorbing effect.

Lubricants also provide a cooling effect on equipment, limiting the stress imposed by heat on their metal parts. They also serve to seal out dirt and other foreign matter from internal working parts and depending on the application, are even used to transmit power hydraulically. So you see, there's much more to lubrication than packing a zirk fitting with grease or filling up an oil cup.

Types of Lubricants

Lubricants are classified by state as liquid, semi-solid, solid or gaseous. Liquid lubricants are derived from hard (stearin) or soft (lard) animal or fish fats; vegetable plants and seeds such as cotton and soybeans; refined from crude as mineral oil, or synthetically produced by man using non-petroleum-based substances in con junction with some combination of these. Oils derived from petroleum are either asphaltic based, containing heavy tar-like materials; paraffin based, possessing large amounts of wax-like materials; napthenic based, containing large amounts of naptha; or mixed-base, bearing some or all of these components.

When thickeners (soaps) are added to oils, they become semi-solid greases. Greases range from hard to soft and are recognized by their pumpability, resistance to dissolving in water, ability to maintain their consistency and their melting points. Solid lubricants are formed metallic or chemical compounds used in applications at temperatures outside of the effective operating ranges of oils and greases. Like solid lubricants, lubricating gasses are used in circumstances that would render the other types ineffectual.

Properties of Oil

Aside from color, thickness and smell, there's no reliable way to distinguish one oil from another, depending only on the human senses to evaluate them. But there are important distinctions that must be considered when matching an oil to its intended application. How hot will it get? Will it be exposed to moisture? How much air (oxygen) will it contact? These questions, and many more, must be answered in order to make the proper choice.

The most commonly considered properties are the oil's flash point (temperature at which its vapor will sustain a flame), fire point (temperature at which its vapor will sustain a flame), viscosity (how thick it is), and pour point (temperature at which it ceases to flow). These properties, and many others, are often engineered into the oil to meet pre-determined specifications through the use of additives which enhance their performance.

A few such additives are antifoam agents which break up air bubbles that form when oil is circulated, detergents that hold dirt particles in suspension and keep them from depositing on metal surfaces, and antioxidants which prevent acid formation resulting from the oxidation of oils. Additives are also used to increase an oil's slipperiness, its ability to remain separated from water and to reduce the effect of temperature on its viscosity. The fact is oil can be treated to meet almost any requirement.

Plant Applications

Knowing your equipment's operating constraints is critical to the selection of lubricants. Operating speeds, temperatures and pressures, hours of operation, loading and even environmental conditions need to be taken into account. Refrigeration systems, for example, operate over wide temperature ranges and require wax and moisture-free oils to avoid problems of ice and acid formation. Electric motors come in sizes from fractional to thousands of horsepower, and have lubrication requirements running the gamut, from good quality machine oils to heavy greases.

Gear oils are used in applications where film strength is a determining factor when considering the pressure or loading involved in mechanical power transmission. Where the speed of rotating parts is of prime importance, low viscosity spindle oils are often specified. If you have any doubt as to which lubricant is appropriate for a particular plant application, it's best to contact the manufacturer for a recommendation.

Lubricant Storage

Most physical plants don't require extremely large quantities of lubricants to be maintained for their daily use, but due to the diversity of their operations they often stock many different kinds. And as most oil and grease is packaged in units that are usually not completely consumed in one application, they pose a special storage problem for the operating engineer. Add to that the slippery nature of the products and the problem becomes a hazard. Compound that with their flammability and the hazard becomes an accident waiting to happen. But accidents need not occur. All that's necessary to avoid the downside of the lubricant equals-disaster equation is to use a little common sense during their handling and storage. Here are a few thoughts on the subject you might wish to consider when evaluating your plant's procedures:

- store all lubricants in well-ventilated areas
- supply dry filtered air to storage rooms
- install fire protection equipment in storage rooms
- keep all open containers tightly covered
- label all the contents of all open containers
- discard old lubricants in approved containers
- never use contaminated lubricants

Appendix C—Special Operating Concerns

- electrically ground all metal drums
- store all lubricants safely above the floor
- never leave oily rags laying about
- keep lubricants separate from potential contaminants
- clean up oil spills when they happen
- maintain lubricant accessories in good repair
- use proper techniques when handling containers
- keep sharp objects away from containers
- allow no open flames in storage rooms
- discard empty and damaged containers
- protect lubricants against moisture exposure
- maintain storeroom within recommended temperatures
- keep records of lubricants stored and used

Appendix D

Stationary Engineer Duties & Traits

Operates and maintains mechanical equipment, such as compressors, generators, and steam boilers, to provide for utilities such as light, heat, and power for refrigeration, air conditioning, and other services for the hospital.

Performs routine maintenance, lubrication, cleaning of mechanical equipment.

Inspects and checks the operation of equipment periodically to see that safety regulations are met and efficient operations are maintained.

Starts and stops equipment, observing meters and gauges and manipulating valves and other controls to regulate flow of water and fuel; adjusts fuel feed, velocity, and location and volume of air in furnaces.

Computes air combustion efficiency.

Checks safety shut down equipment.

Tests water condition and adds chemical treatment to maintain proper limits and to prevent deposits.

Removes water samples and performs chemical tests to determine water hardness and properly applies chemical softeners, as needed.

Checks equipment while in operation and makes minor repairs, replaces bearings, and places and adjusts piston clearance, as necessary.

Demonstrates responsibility for oiling and greasing moving mechanisms, maintaining proper water conditions and cleaning related equipment.

Observes regulations imposed by local, state or other authorities.

Demonstrates the ability to replace worn and defective parts on equipment.

Maintains and repairs refrigeration equipment and hot water heaters to ensure maximum operating efficiency; adjusting feed valves and pump speeds, as required.

Accurately prepares daily reports.

Performs routine maintenance, lubrication and cleaning of equipment.

Maintains equipment in a clean and safe operating condition.

Is familiar with and checks chemical supplies and materials used by the department and submits requisitions for reorder.

Disassembles and cleans mechanical/electrical devices.

Starts, stops, adjusts and regulates mechanical equipment such as engines, compressors, generators, pumps and steam boilers.

Checks water levels in boilers as well as domestic hot and cold water tanks.

Observes meters, recorders, pressure and draft gauges and manipulates valves to bring readings within specified requirements.

Checks all mechanical rooms during the respective shift, recording readings in a logbook or on other forms.

Inspects equipment while in operation and makes minor repairs and/or adjustments when necessary.

Appendix D—Stationary Engineer—Duties and Traits

Starts, stops, and demonstrates the ability to adjust and regulate equipment.

Observes steam-flow meters, carbon dioxide recorders, pressure and draft gauges and manipulates valves to bring readings within the specified requirements.

Understands and participates in cost control activities by monitoring utilization of supplies, equipment and expenses.

Evaluates and sets standards for equipment usage and monitors utilization of equipment/supplies.

Assists in the maintenance of all capital and miscellaneous equipment; always makes sure that equipment is in safe and proper working order.

Maintains equipment and supplies properly; schedules repairs and orders supplies as necessary; demonstrates cost consciousness in using and ordering supplies.

Receives stock supplies; dates and labels items as required; verifies that proper supplies have been received in the ordered quantity.

Demonstrates knowledge of and rationale for safe use of equipment; reports any equipment malfunctions and orders service, as necessary.

Knowledgeable of process for the containment of costs and conservation of supplies and equipment.

Uses forms appropriate to the task; demonstrates a high degree of accuracy and attention to detail.

Knowledge of hazardous materials handling according to proper policy guidelines.

Operates emergency generators for test purposes.

Disassembles and cleans equipment, checks shut-down equipment.

Performs all duties in an independent manner with little or no need for direct supervision.

Orders all necessary equipment and supplies; demonstrates a cost conscious attitude; regularly inventories to ensure supplies are always on hand.

Maintains a safe environment in accordance with policies and safety regulations.

Demonstrates a proactive attitude and seeks to remedy situations before an accident or mistake is made; assists co-workers that exhibit insufficient knowledge or ability to complete the assigned tasks/duties.

Assists in the development of policies, procedures and forms applicable to the department and submits them for approval and implementation.

Makes suggestions for changes in policies or procedures that would improve efficiency.

Uses reference material to ensure accuracy.

Prepares for inspections and special maintenance procedures.

Has a complete understanding of fire and safety procedures as outlined in the Fire and Safety manual; knows personal role in case of a fire during the assigned shift.

Demonstrates good judgment in routine answering and referral of phone calls; advises appropriate personnel of situations requiring follow-up attention.

Recognizes unsafe acts or conditions and takes action immediately; implements action to eliminate future problems of the same sort.

Assesses the physical environment and reports needed repairs or alterations to the appropriate personnel.

Follows safety procedures as outlined in the departmental procedure manual at all times.

Demonstrates the ability to relate well to all levels of staff.

Appendix D—Stationary Engineer—Duties and Traits

Demonstrates good verbal and written communication skills; documents and reports clearly and concisely.

Demonstrates effectiveness in identifying future needs and problem areas of the department; developing workable solutions; following through on solutions.

Assists the supervisor in planning and implementing new programs with special attention to deadlines and other constraints.

Prepares a daily work schedule for self, scheduling appropriate amounts of time for preparation and completion of tasks; coordinates schedule with other staff members.

Demonstrates an ability to perform in an accurate and precise manner in crisis and emergency situations when time is critical.

Familiar with stocking, inventorying, clerical duties, ordering, cleaning and special projects procedures.

Demonstrates a general knowledge of boilers, pumps, gauges, heating units, hand tools, refrigeration and ventilation equipment, plumbing fixtures and steam distribution systems.

Demonstrates a working knowledge of electricity, water treatment and steam trap operation.

Obtains and maintains state licensure, as required.

Demonstrates a thorough knowledge of safety policies and procedures; utilizes proper body mechanics; observes and follows good safety practices at all times.

Demonstrates an active interest in improving current level of skills and knowledge by independently participating in facility learning experience.

Assists in the development of manuals, guidelines, records and other informational materials for the department.

Uses policy and procedures manuals, office procedures manual and other reference materials as necessary to ensure proper course of action.

Assists in the orientation and skill development of new staff members, as required.

Demonstrates the ability to maintain all equipment in the assigned work area assuring that it is in safe and proper working order.

Demonstrates understanding of emergency procedures, fire drills, evacuation procedures and internal/external disaster plans and procedures.

Conducts periodic mechanical room inspections; ensures that all standards are maintained.

Maintains the proper operation of the emergency electrical generating systems.

Demonstrates an ability to assess a situation from a variety of perspectives, consider several alternatives, and choose an appropriate course of action.

Recognizes his/her role in the department and how it relates to the overall function.

Demonstrates an ability to work effectively with a variety of co-workers regardless of varied racial, ethnic, and sociological backgrounds.

Demonstrates effective social and communicative ability; conveys instructions, directions and information appropriately.

Displays an unhurried, competent manner; demonstrates the ability to remain friendly and cooperative during all working conditions.

Coordinates work to achieve maximum productivity and efficiency.

Strives to meet deadlines in special projects and daily responsibilities.

Demonstrates the ability to recognize, establish and deal with priorities promptly.

Is capable of preparing monthly statistics of departmental activities in order to better plan and prepare the schedule of operation.

Assists others in need, fills in free time with clerical duties, special projects, cleaning, etc.

Prepares and submits monthly statistics of departmental activities to the supervisor in order to better plan and prepare the schedule of operation for the entire department.

Seeks information necessary for accurate completion of job duties; uses reference material to ensure accuracy.

Demonstrates an ability to learn and adapt to changes in function, management style or department routines.

Maintains work area in a manner conductive to efficiency and safety and cleans and organizes department on own initiative.

Ensures supplies are maintained according to established guidelines and monitors usage to determine need for a change in established levels.

Works independently. Carries out assignments promptly.

Demonstrates responsibility through responsiveness to others; competently follows up on matters requiring additional attention.

Demonstrates reliability in the conscientious and complete manner in which work assignments are performed.

Always completes documentation and paperwork in a timely manner.

Keeps up to date on all procedures with regard to duties—knowing when, where, and how to do the assigned job.

Performs a variety of duties in and around the buildings of the complex to assist the tradesmen and other workmen in completing their tasks; repairs as well as installation of new service or projects as needed.

Picks up various supplies, parts, and building materials and delivers to appropriate location as required.

Inspects, cleans and rebuilds traps and strainers.

Services and repairs exhaust fans, steam leaks, fan motors and fan belts.

Solders, brazes and welds material.

Repairs gas leaks, cleans electronic filters, inspects and safety checks equipment.

Checks pump and motor operation, components on air handling systems, and turns off systems for energy conservation.

Replaces fuses and resets circuit breakers.

Performs filter changes in air handling equipment, picks up parts and supplies in emergencies and cleans work area and shop area.

Installs equipment of various types without supervision.

Capable of working within set standards and limits.

Repairs machinery and mechanical equipment in accordance with diagrams, sketches, operation manuals and manufacturers' specifications.

Demonstrates the ability to observe mechanical devices (pumps, engines, motors, etc.) in operation and listen to their sounds to locate causes of trouble.

Dismantles devices to gain access to and remove, repair and replace defective parts.

Adjusts functional parts of devices and control instruments or installs special functional and structural parts, as required.

Appendix D—Stationary Engineer—Duties and Traits

Lubricates or cleans parts and starts devices to test their performance.

Demonstrates the ability to set up and operate lathe, drill press, grinder, and other metal-working tools to make and repair parts.

Keeps record cards for equipment indicating type, model number, date of installation and extent of servicing.

Maintains high degree of accuracy when making adjustments or aligning equipment.

Receives stock supplies; dates and labels items as required; verifies that proper supplies have been received in the ordered quantity.

Demonstrates knowledge of and rationale for safe use of equipment; reports any equipment malfunctions and orders service.

Checks emergency equipment and ensures staff knowledge of proper use.

Always uses forms appropriate to the task; demonstrates a high degree of accuracy and attention to detail at all times.

Assists in the maintenance of all capital and miscellaneous equipment; always ensures that equipment is in safe and proper working order.

Demonstrates the ability to plan ahead in cases of emergencies.

Strives to make the best use of time during the assignment shift through careful coordination of daily tasks; reduces non-essential interruptions to an absolute minimum in order to complete priority tasks first.

Educational qualifications should include graduation from trade or vocational school.

Demonstrates the ability to read blueprints or drawings, a thorough knowledge of hand and power tools, precision-measuring and testing instruments.

Demonstrates numerical ability needed to make calculations for installation and repair of equipment.

Demonstrates the ability to repair or replace all piping (water or steam), plumbing fixtures, or other maintenance requirements.

Demonstrates a basic knowledge of electricity, plumbing, carpentry and painting.

Keeps tools and equipment clean and lubricated; repairs broken tools and equipment.

Demonstrates understanding of emergency procedures, fire drills, evacuation procedures and internal/external disaster plans and procedures.

Identifies and actively pursues a planned program for professional growth through night classes, reading textbooks, consultations with others, facility learning experiences, etc.; demonstrates a sincere interest in improving current level of skills and knowledge.

Maintains active membership in local trade associations in order to expand personal knowledge base and develop professional contacts.

Demonstrates the ability to work alone and make decisions necessary to complete the project(s).

Remains alert and demonstrates the ability to make decisions regarding the trouble-shooting of problems.

Demonstrates good judgment in evaluating malfunctioning conditions against verifiable or judgmental criteria as shown in blueprints or installation instructions.

Participates in the evaluation of new products, equipment, procedures and maintenance routines.

Recognizes the role of a maintenance mechanic in the department and how it relates to the overall function.

Appendix D—Stationary Engineer—Duties and Traits

Recognizes unsafe acts or conditions and takes action immediately; continuously monitors unsafe acts or conditions until satisfied that appropriate action has been taken; implements action to eliminate future problems of the same sort.

Foresees potential problem situations; intervenes to offset adverse impact; demonstrates a proactive attitude in maintenance of equipment.

Demonstrates competence in the day-to-day scheduling of duties by completing all priority and related tasks on time with little additional assistance required from others.

Never makes hasty decisions; obtains and analyzes all pertinent information available in order to make the most informed decision based on factual and objective data.

Makes suggestions as to ways to complete assignments easier, quicker and at less cost.

Maintains work area in a manner conductive to efficiency and safety; cleans and organizes department on own initiative.

Demonstrates a proactive attitude and seeks to remedy situations before an accident or mistake is made; assists co-workers who exhibit insufficient knowledge or ability to complete their assigned tasks/duties.

Ensures that supplies are maintained according to established guidelines; monitors usage to determine need for a change in established levels.

Reports all breakdowns of equipment; works to repair same as quickly as possible.

Demonstrates reliability in the conscientious and complete manner in which work assignments are performed.

When requested, is always willing to adjust personal schedule to complete work load, prepare for an inspection or special maintenance procedure.

Demonstrates willingness to learn, attend seminars and schools that are beneficial to self improvement.

Performs as a versatile individual with good mechanical abilities; works well in industrial environments.

Demonstrates ability to work in an environment subject to noise from machinery; follows safety procedures carefully to avoid electrical shock or burns from heated equipment.

Demonstrates ability to climb, balance, stoop, kneel and crouch, as necessary, to make repairs or installations.

Performs assigned work orders and files starting and completion times and list of materials used.

Performs mechanical duties based upon work order requests or PM cards.

Performs readings on all electrical and mechanical equipment that is essential for daily operations.

Performs water tests according to guidelines set by water treatment company and document all findings in water treatment log book; mixes and adds chemicals according to guidelines to maintain proper readings.

Prepares lists of parts or equipment needed to perform prescribed duties.

Requests parts and materials and verifies description, quality, quantity and other detailed information.

Initiates necessary action to test fire alarm system; maintain sprinkler system records.

Makes rounds and inspects fire valves to assure they are opened and in locked positions; documents all information.

Provides advice and guidelines to employees related to policy interpre-

Appendix D—Stationary Engineer—Duties and Traits

tations and related matters; ensures compliance with local building codes and other applicable regulations.

Removes defective parts and installs new parts, completes safety and performance checks before returning to service.

Prepares a history and PM card on each piece of equipment received which requires a preventive maintenance program.

Read blueprints, manufacturer's specifications, and instructions regarding work assignments.

Analyze all types of drawings and wiring diagrams to determine material or replacement needs and to select a logical approach to troubleshooting and repair problems.

Maintain record of critical adjustments, repairs, and spare parts used and runs performance checks on operating equipment.

Observe mechanical devices, such as pumps, engines, air compressors, laboratory equipment, boilers, etc., in operation and locate causes of trouble.

Adjust functional parts of devices and control instruments or install special functional and structural parts.

Perform tests to check the operation or performance of repaired mechanical devices using meters, test equipment, and precision measuring devices.

Make minor repairs to electrical fixtures, plumbing, pipe fittings, etc. and required.

Operates various machines, such as lathe, drill press, grinders, and other metal-working tools or equipment, to make or repair parts.

Inspects and operates refrigerating machinery and cooler and freezer units to determine that all are functioning properly.

Maintain and repair refrigeration and air conditioning systems; read thermometers and gauges, record readings, adjust pipeline valves, and regulate refrigerant and suction valves.

Check joints for leaks, trace electrical control systems for loose or shorted wires, and test operation of thermostats with electrical testing equipment.

Resolder joints or repacks, replace valves, dismantle compressor or replace worn parts, and dehydrate system.

Installs refrigeration machinery and air conditioning equipment.

Mounts compressor and condenser units, installs evaporator units and tubing, and joins them to complete circuit for the refrigerator.

Install and connect independently operated water coolers and household size refrigerators using small hand tools and following manufacturer's instructions.

Inspects and tests electrical lighting, signal, communication, emergency electrical generating equipment, engines, motors, power circuits and equipment, etc.

Isolates defects in wiring, switches motors, and other electrical equipment using various electrical testing equipment.

Dismantles electrical machinery and replace various defective mechanical parts and related electrical parts, assembling components according to diagrams.

Checks clearances of moving parts with precision gauges and restores electrical connections to complete circuits.

Install new wiring and electrical machinery by running wire and conduit and fastening fixtures, switches, and outlet boxed in position, and then making connections to complete circuits.

Inspect, repair, and maintain piping systems, plumbing fixtures, outlets, and heating, water, and gas drainage systems, etc.

Replace defective pipe and tubing by cutting, bending, threading and assembling pipe or tubing.

Appendix E

Universal Task Instructions

(Additional tasks can be fabricated by referencing equipment manuals, canned data bases, plant operating records and personal familiarity with the subject matter being addressed.)

EQUIPMENT/SYSTEM CHECKS

Air Outlets
Inspect air outlets at rear or top of machine for blockage

Battery Fluid Level
Check each cell of battery for proper fluid level and add distilled water if required.

Door Hardware
Repair, adjust and lube as necessary. Make certain that latch holds the door closed security and that the latch works freely.

Dry Sprinkler System
The air or nitrogen pressure on each dry-pipe system should be checked at least once a week and maintained as per manufacturers instructions. All leakage resulting in pressure loss should be repaired.
Record gas and water pressures.
Insure that each head is free from any hanging items (wires, tape, cords, etc). Items stacked below the heads must be a minimum of 12 inches below the heads.

Fan
Check for vibration and noise.

Fire Standpipes
Charge and activate fire standpipes to check for proper operation.

Gas Piping And Valves
Check at critical points for deposits. Check gas expansion tube for plugs.

Gas Solenoid Valves
Check condition of plunger and diaphragm and replace if necessary.

General Equipment Check
Visually inspect the general condition of equipment, paying close attention to possible safety hazards and note findings.

Hot Water Temperature
Check water temperature at thermometer on unit and compare to a fixture in another location; note both.

Motors
Clean and inspect all drive, pump and blower motors. Check all related connections-electric, water, air and others.

Notification Of Supervisor
Locate or call the area supervisor and apprise him/her of the condition and status of this equipment and inform all users of equipment status.

Pump Suction & Discharge
Insure that suction and discharge valves are in fully open position. Observe and note suction and discharge pressures.

Spray Nozzles
Check for tightness and alignment.

Switches, Wiring and Connections
Operate and visually inspect for proper function.

Thermostats
Ascertain that they cycle at proper temps.

Appendix E—Universal Task Instructions

Transfer Switch
Check mechanical interlock linkage to prevent both switches from closing at the same time.
Test operation of switch and verify control and timer operation.
Inspect panel doors, hinges and latches for proper fit.

Vacuum Breaker
If water is seen leaking from top of vacuum breaker, make necessary repairs.
Make certain that supply pressure is adequate.

Water Flow
Check to make certain that all valves are open fully and water flow to unit is unrestricted.

Wiring
Inspect for loose or frayed wiring and poor connections.

EXTERIOR WORK

Solid Waste Management
Collect/dispose of refuse and solid waste. (Includes incinerator operations, duty, and trash can collections.)

INSPECTION

Abnormal Conditions
Inspect the area for any abnormal conditions, paying attention to the appearance, integrity and safety of the unit being inspected.
Items to be included are paint, flooring coverings, ceilings, plumbing, lighting, electrical outlets, and miscellaneous equipment.
Correct any abnormal condition or initiate work order if correction is beyond the inspectors capability.

Ash Conveyor System
Drain ash quench tank and clean.
Inspect all below water components.
Grease and inspect lower and upper chain bearings.

Check drive assembly for proper alignment.
Adjust chain tension.
Refill tank.

Bag House & Air Lock
Open and inspect bag house for excess lime build-up.
Grease air lock rotor bearings and lube drive chains.

Boiler Sight Glass
Inspect sight glass for signs of leakage.
Repack ends of sight glass.

Boiler Tubes
Inspect each boiler tube for warping, deterioration and material accumulation.

Burner Assembly
Dismantle and clean pilot burner assembly.
Inspect conditions & connections of ignition leads to burners and replace as necessary.
Clean face of flame detection tubes as needed.

Coupling
Visually inspect to insure that motor & pump shafts are aligned. (Adjust using shim).
Insure the two halves of the coupling are snug against the rubber sleeve and tighten the set screws.

Deaerating Tank
Install temporary feedline from condensate tank to boiler feedwater pump.
Remove DA tank from service, secure steam to DA tank and give tank time to cool.
Open manhole covers and flush tank with water hose.
Inspect diffusers in top section of tank.
Repack valves and sight glasses as needed.
Reassemble tank, fill with feedwater and start pumps.
Secure and remove temporary feedline from condensate tank.

Door Closures

Check to see that the closure will hold the door closed securely and fully open.
Make certain sequential closing function works.
Check to see that the gasket is supple and sealing properly.
Check door gasket for cracks or brittleness.
Clean door surface, door frame surface and gasket surface with solvent making certain that surfaces are free of foreign matter.
Make sure door guides are straight and fastened securely.

Drainage System

Inspect trench drains, drain basins, grates, leaders, downspouts, and all support brackets for leaks, damage or distress.
Repair system as necessary.

Heat Exchangers

Inspect for proper operation, leaks, and corrosion.

Hydraulic System

Check hydraulic fluid level and record.
Observe unit for evidence of leakage while operating pump.

Induced Draft Fan

Inspect induced draft fan dampers and fan blades for soot accumulation and clean unit.
Check and tighten fan hub set screws and bearing hold down, foundation and motor mounting bolts.
Examine all surfaces for erosion and excessive wear.
Check belt tension.
Balance fan, if necessary.

Lamp Holder And Lamp

Inspect the lamp for damage or discoloration and replace if necessary.
Check for burned socket contacts.
Make sure lamp is firmly seated in its socket.
If wires are frayed or insulation is worn, replace wires.
Install lamp holder assembly making sure the wiring doesn't interfere with its operation.

Light Fixtures, Electrical Controls And Panels
Inspect panel boards, circuit breakers, contactors, relays, switches, motors, wiring terminations and grounding for proper operation and condition.

Mechanical Area
Inspect the area for abnormal conditions, damaged equipment and/or safety hazards.
Correct problems found or write a work order and turn it into your supervisor.
Record all readings taken in the appropriate logs.

Pump Packing
Run pump with water flow normal. (A slight leak of a few drops a minute through the packing is desirable to lubricate and cool the shaft.) If leakage is less than this amount, loosen the packing nut; if more, tighten slightly.

Safety Valve
Inspect for accumulations of rust, scale or other foreign substances which would prevent free operation of the device.
Ascertain that all discharge pipes are free and clear of obstructions.
Test the valve by operating the manual lifting lever. (The lever should move freely and return to the closed position after each operation.) If leakage is evident, try operating the lever several times.)

Fuel Oil Strainers
Set strainer selector lever to put clean strainer in service.
Remove cover from dirty strainer; clean strainer in solvent.
Replace strainer and cover.
Test for leaks.

Fuel Oil Heaters
Secure inlet and outlet valves on heater.
Fill heater with no-sludge solvent, using connections on heater. Circulate solvent through heater using auxiliary gear pump.
Drain heater and operate under normal conditions.
Inspect for leaks.

Appendix E—Universal Task Instructions

Fuel Tank
Drain fuel tank until clean fuel flows from drain plug at bottom of fuel tank.

Hinges, Locks & Weather-stripping
Apply machine oil to all door hinges, hood hinges, tail gate hinges, etc.
Apply weather-stripping lubricant to all rubber weather-stripping and stops.
Apply graphite to all locks.

Incinerator Burner
Clean fan and fire-eye controller.
Adjust forced draft fan louver, clean and lubricate motor using a light machine oil.
Test flame failure devices.

Incinerator Combustion Chamber
Before entering incinerator combustion chamber, make certain they have cooled and all ashes have been removed.
(DO NOT SPRAY WATER ON HOT REFRACTORY, THIS COULD CAUSE DAMAGE)
Inspect refractory, thermocouples and clean burnercones.

Incinerator Maintenance
Open air valves.
Check hydraulic fluid level.
Check ash conveyor float.
Verify all control set points.
Check loader slide carriage, cable and limit switches.
Check boiler water level and feedwater pressure.
Inspect and clean fire door shroud around hydraulic cylinder shafts.
Inspect fire door hydraulic cylinder mounting bolt alignment.
Inspect and secure all air tube mounting brackets.
Check operation of ID fan dampers and draft, lime feeder, auger system, water sprayers, and bag house differential pressure;
Inspect & adjust transfer ram limit switches.
Inspect transfer ram wipers.
Inspect ash conveyor drag chain for tension; adjust as required.
Verify proper operation of the ash conveyor on timed or continuous cycle.

Inspect operation of combustion air fan for noise, vibration, and function monitoring.
Check operation of all dampers, linkage assemblies and modulating motors.

Motors
Remover motor end bells, remove rotor, blow internal parts out with compressed air.
Inspect bearings.
Reassemble motor, operate under normal loading conditions.
Using an AMP Meter compare motor amps to name plate data.

Printed Circuit Boards
Clean all the solid state cards in the card file and reseat cards tightly into their slots.

Pumps Packing
Operate pumps at normal pressure and oil at normal temperature and inspect pump shaft packing.
(To replace packing secure pump, remove packing gland and old packing.
Make a note of the number of rings of packing removed and replace the same number).
Operate the pump and adjust packing leakage.

Refrigeration unit
Inspect/clean evaporator and condenser coils.
Clean and check components for proper operation.
Clean out compartments recharge if needed.
Replace panels.
Note any abnormal conditions found.

Restore Operation
Restore all connections and utilities to operational status

Sealant Systems
Remove existing sealant systems in all joint systems.
Replace with new sealant material.

Spark Plug(s)
Replace all spark plugs. (Gap in accordance with maintenance manual.)

OPERATION

Automatic Door
After door is open, stand in doorway with one foot on each side of the door sill; (door should remain open).
If door closes, repair or replace the safety apparatus.
Walk through the door and off the safety mat; turn around and step back on the safety mat.
Wait for someone to go through the door way; the door should not open in your direction until you step out of the door swing, i.e. off the safety mat.

Bearings
Using high quality lubricant, lubricate all bearings thoroughly and wipe off any excess.

Boiler Shutdown
Raise water level to within one inch of top of sight glass.
Let boiler set for about two days to lower pressure and to cool.
Close fuel valves and all pressure valves to boiler.
Secure valves with tags, dated and signed, not to be opened for any reason.
Main steam stop and bottom blow down valves should be chained and pad locked closed.

Condenser Water Flow
Valve should be open enough so there is free flow of water through.
should be cool enough to hold hand on without discomfort.

Electrical Movement And Safety Precautions
Exercise caution during maintenance operations and you will reduce your chances of getting electrically shocked.
Do not wear rings or watches when working and be sure to use tools with insulated handles.

Electrical Power
Restore power at disconnect

Elevator
Respond to call. Correct the problem if possible.
Insure that the elevator is not on manual operation.
If problem is beyond your capability to correct, contact the appropriate elevator contractor to make the necessary repairs.
While door is closing tap the underside of safety edge at the opposite side from the pressure switch to insure that the safety switch is functioning properly.

Engine Cooling System
Remove blower housing and clean air screen and engine cooling fins with air hose

Equipment Observation
Run unit and check function of indicator lights, manual cycle indicator, selector buttons, and buzzer.
Operate re-set button.

Fan Bearings
If fan has oil bath bearings, drain casing and refill with proper grade oil according to specifications then check for leaks.

Fixtures
Check drain valve for proper operation.
Check floor drain or drain pipe for unimpeded flow, remove faucet aerator and clean with bleach.
Repair surface cracks and chips.
Check supply valves for leaks and proper operation and repair as necessary.
Clean trays.

Gas Pressure
Reconnect cylinders and pressurize system.

Incinerator Start-Up
Check that all doors are closed.
Check water level in ash conveyor.

Verify city water.
Clean rear of loader ram, add lime if necessary.
Check air pressure.
Start system, manually run ash conveyor.
Inspect & clean all cooling fins and screens on electric motors using air hose.
Inspect all motors for bad bearings while operating.
Check water level in ash quench tank and verify operation of the tank flood valve.
Remove debris from and inspect ash conveyor shaft and cogs.

Incinerator Operation
Check ID damper operation and draft.
Check all temperatures on environmental panel.
Check operation of bag house and boiler soot blowers.
Observe operation of upper and lower dampers.
Check for water, steam leaks and hydraulic leaks.
Blowdown boiler and sight glass as needed.
Observe several load cycles.

Mechanical Space
Conduct mechanical room inspections of equipment and utility systems to check for unusual conditions or to insure equipment/systems are operating within specified limits.
Record readings as directed by shop manager.

Steam Lines To Recording Equipment
Close valves to recording equipment and open pressure equalizing valve.
Open Blowdown valves slowly until live steam comes out of Blowdown line.
Close valves and let system fill with condensate before opening service valves to recorders.

Sump Pump
Watch for the following things:
a) Sticking:
    See that the unit does not stand idle for very long periods. If necessary start pump manually to see that the shaft is free.
b) Automatic equipment:

Check frequency of starts and stops to see that unit is properly regulated.

Check contacts of switches cleaning them and applying Vaseline if they show signs of burning. (If contacts are badly burnt replace the switch before serious consequences result).

c) Motor:
See that the motor comes up to speed quickly, maintains constant rotation rate, and does not spark profusely.
Keep the interior and exterior of motor and automatic apparatus free from moisture, oil, and dirt.

d) Basin cleaning:
Occasionally run the pump until the basin is drained of water, break the electric circuit, disconnect the discharge pipe at the union.
Clean out sludge or foreign matter accumulated in the basin and on the strainer before returning the pump to service.
Lubricate motor and pump.

Turn Off Electric Supply
Shut off all power at main disconnect

Unit Function
Operate unit through entire cycle and observe for proper function.

TESTING/CALIBRATION

Antifreeze
Test the strength of the antifreeze using a hydrometer.
Add antifreeze to protect to prescribed temperature.

Battery Cell Voltage
Using a cell analyzer, start with red (+) lead on the positive post of the battery, insert the black lead (–) into the first batt cell until it touches only the electrolyte fluid, meter should read 2 volts.
Continue this procedure until all cells have been checked.
A variation of 1/2 v or more indicates a defective cell.
Battery should be charged.

(When cleaning batteries, eye protection must be worn).
Check electrolyte level and take hydrometer reading of all cells, the

Appendix E—Universal Task Instructions

level of electrolyte should be well above the cell plates.
Fill with distilled water to the specified level. (Do not overfill). Inspect and test auto charging receptacle.
Clean dirty contacts. (All contacts that appear to have been arcing must be cleaned to a smooth finish).
Check polarity of charging plug.
Measure connection with voltmeter.

Electrical receptacles
Visually inspect for defects, tension, polarity and ground.
Replace substandard or defective receptacles and note location and replacement dates.

Fire Alarm Test
Call Telephone Operator and announce the test.
Activate a pull station or smoke Detector to make the test.
Record results.
Make necessary adjustments or repairs to place system in an operational status.

Manufacturers Instructions Notation
Follow manufacturers instructions for inspection, cleaning and lubrication of the equipment.

Mechanical Room Equipment
Test equipment to insure it is functioning properly.
Correct any deficiencies and record findings or initiate a work order for corrective action.

Operating Controls
Using appropriate measuring devices or values calibrate controls

Pressure And Vacuum Switches
Make certain that all switches operate within their correct range.

Safety Valves
Pull safety valve lever (valves should pop to an open position) Release pressure on lever. (valve should snap shut without chattering or leaking).

Sensors
Remove sensor cover.
Remove sensor board and clean light sensor units and mirrors in the housing.
Reassemble unit and test.

Smoke Detectors
Test 2 smoke detectors (one on each zone to verify operation).
Alternate the 2 detectors tested each time and keep a record of this so all detectors will eventually be tested.
When second detector goes into alarm the horns will begin to pulse.
Check and verify that system has been reset and is in normal operating mode.
Include all devices that are part of the system.

Transmission
(While motors is running) oil level should read between the two marks.
Adjust level as necessary

Steam Turbine Test
Start turbine auxiliary equipment by cutting in a very small amount of steam into turbine and turbine case.
Start turbine shaft rotating very slowly.
Slowly bring turbine full speed.
Operate for one hour at full speed before testing overspeed trip.
Manually raise speed until overspeed trips closing the main steam valve.
Monitor the RPM's and record same.

APPENDIX F

LIST OF TABLES

Air
1. BTU Required for Heating Air
2. Composition of Air
3. Atmospheric Pressure per Square Inch
4. Approximate Air Needs of Pneumatic Tools
5. Average Absolute Atmospheric Pressure

Construction
6. Building Design Loads
7. Concrete Curing Methods
8. Concrete for Walls
9. Earth Excavation Factors
10. Lumber Sizes in Inches

Conversions
11. Weight
12. Temperature Conversion
13. Length and Area
14. Horsepower Equivalent
15. Approximate Metric Equivalents
16. Miscellaneous Measures

Electric
17. Appliance Energy Requirements
18. Alternating Current Calculations
19. Kilowatt Conversion Factors
20. Effects of Electric Current on Humans
21. Equivalent Electrical Units

22. United States Power Characteristics
23. Induction Motor Synchronous Speeds

Formulae
24. Geometric Formulas

Heat
25. Heat Equivalents
26. Heat Generated by Appliances
27. Heat Loss from Hot Water Piping
28. Heat Loss from Bare Steel Piping
29. Specific Heat of Common Substances
30. Heat Content of Common Fuels
31. Typical Fuel Oil Analysis

Miscellaneous
32. Cylindrical Tank Capacity
33. Metric Capacity
34. Metric Length
35. Metric Weight
36. Weights of Metals
37. Ultimate Bending Strength
38. Recommended Lighting Levels

Steam
39. Steam Pressure Temperature Relationship
40. Steam Trap Selection

Temperature
41. Color Scale of Temperature
42. Centigrade/Fahrenheit Scale
43. Melting Points of Common Substances
44. Temperatures of Waste Heat Gases

Water
45. Water Equivalents
46. Supply Line Sizes for Common Fixtures
47. Pressure of Water
48. Water Requirements of Common Fixtures
49. Blowdown Flow Rates-3 Inch Pipe
50. Tank Capacities Per Foot of Depth
51. Efficiency Loss Due to Scale

AIR

Table 1. Btu Required For Heating Air
(1 cubic foot of air)

External Temp	\multicolumn{4}{c}{Temperature of air in room}			
	60	70	80	90
0	1.234	1.439	1.645	1.851
10	1.007	1.208	1.409	1.611
20	0.787	0.984	1.181	1.378
30	0.578	0.770	0.963	1.155
40	0.376	0.564	0.752	0.940
50	0.184	0.367	0.551	0.735
60		0.179	0.359	0.538
70			0.175	0.350

Table 2. Composition Of Air

Nitrogen	78%
Oxygen	21%
Argon	0.96%
Carbon Dioxide & other gases	0.04%

Table 3. Atmospheric Pressure Per Square Inch

Barometer (ins. of mercury)	Pressure (lbs. per sq. in)
28.00	13.75
28.25	13.88
28.50	14.00
28.75	14.12
29.00	14.24
29.25	14.37
29.50	14.49
29.75	14.61
29.92	14.69
30.00	14.74
30.25	14.86
30.50	14.98
30.75	15.10
31.00	15.23

Table 4. Approximate Air Needs Of Pneumatic Tools, CPM

Grinders,   6- and 8-in. diameter wheels 50
            2-and 2 1/2-in. diameter wheels 14-20

File and burr machines 18
Rotary sanders, 9-in. diameter pads 55
Sand rammers and tampers:

    1 × 4-in. cylinder 25
    1 1/4 × 5-in. cylinder 28
    1 1/2 × 6-in. cylinder 39

Chipping hammers,   10-13 lb 28-30
                    2-4 lb 12

Nut setters to 5/16 in., 8 lb. 20
    1/2 to 3/4 in., 18 lb 30

Paint spray 2-20
Plug drills 40-50
Riveters, 3/32-to 1/8-in. rivets 12
Rivet busters 35-39
Steel drills, rotary motors:
    To 1/4 in., weighing 1 1/4-4 lb 18-20
    1/4 to 3/8 in., weighing 6-8 lb 20-40
    1/2 to 3/4 in., weighing 9-14 lb 70
    3/8 to 1 in., weighing 25 lb 80
    Wood borers to 1-in. diameter,
        weighing 14 lb 40

Appendix F—List of Tables

Table 5. Average Absolute Atmospheric Pressure

Altitude in feet reference to sea level	Inches of Mercury (in. HE)	Pounds per sq. in. absolute (Psia)
− 1,000	31.00	15.2
− 500	30.50	15.0
sea level 0	29.92	14.7
+ 500	29.39	14.4
+ 1,000	28.87	1 4.2
+ 1,500	28.33	13.9
+ 2,000	27.B2	13.7
+ 3,000	26.81	13.2
+ 4,000	25.85	12.7
+ 5,000	24.90	12.2
+ 6,000	23.98	11.7
+ 7,000	23.10	11.3
+ 8,000	22.22	10.8
+ 9,000	21.39	10.5
+ 10,000	20.58	10.1

CONSTRUCTION

Table 6. Building Design Loads

Occupancy or Use	Live Load Lbs. per Sq. Ft.
Apartment houses:	
Private apartments	40
Public stairways	100
Assembly halls:	
Fixed seats	60
Movable seats	100
Corridors:	
First Floor	100

(Continued)

Table 6. Building Design Loads (*Continued*)

Occupancy or Use	Live Load Lbs. per Sq. Ft.
(Other floors, same as occupancy served except as indicated.)	
Dining rooms	100
Dwellings	40
Loft buildings	125
Sidewalks	250
Stairways	100

Table 7. Concrete Curing Methods

Method	Advantage	Disadvantage
Wetting	Excellent results if constantly kept wet	Difficult on vertical walls
Straw	Insulator in winter	Can dry out, blow away, or burn
Curing	Easy to apply.	Sprayer needed
Compounds	Inexpensive	Can allow concrete to get too hot
Waterproof Paper	Prevents drying	Cost can be excessive
Plastic Film	Absolutely watertight	Must be weighed down

Table 8. Concrete For Walls
(Per 100 Square Feet Wall)

Thickness	Cubic Feet Required	Cubic Yards Required
4"	33.3	1.24
6"	50.0	1.85
8"	66.7	2.47
10"	83.3	3.09
12"	100.0	3.70

Table 9. Earth Excavation Factors

Depth	Cubic Yards per Square Foot
2"	.006
4"	.012
6"	.018
8"	.025
10"	.031
1'-0"	.037
1'-6"	.056
2'-0"	.074
2'-6"	.093
3'-0"	.111
3'-6"	.130
4'-0"	.148
4'-6"	.167
5'-0"	.185
5'-6"	.204
6'-0"	.222
6'-6"	.241
7'-0"	.259
7'-6"	.278
8'-0"	.296
8'-6"	.314
9'-0"	.332
9'-6"	.350
10'-0"	.369

Table 10. Lumber Sizes in Inches

Nominal	Seasoned
1 × 4	3/4 × 3-1/2
1 × 6	3/4 × 5-1/2
1 × 8	3/4 × 7-1/4
1 × 10	3/4 × 9-1/4
1 × 12	3/4 × 11-1/4
2 × 4	1-1/2 × 3-1/2
2 × 6	1-1/2 × 5-1/2
2 × 8	1-1/2 × 7-1/4
2 × 10	1-1/2 × 9-1/4
2 × 12	1-1/2 × 11-1/4
4 × 4	3-1/2 × 3-1/2
4 × 6	3-1/2 × 5-1/2
4 × 8	3-1/2 × 7-1/4
4 × 10	3-1/2 × 9-1/4
4 × 12	3-1/2 × 11-1/4

CONVERSIONS

Table 11. Weight

1 Pound	0.454 Kilogram
1 Short Ton (2000 lbs)	907 Kilograms
1 Long Ton (2240 lbs)	1016 Kilograms
Gram	15.432 Grains
1 Kilogram	2.205 Pounds
1 Metric Ton	2205 Pounds

Table 12. Temperature Conversions

°F	=	9/5°C + 32
°F	=	°R − 459.58
°K	=	°C + 273.16
°R	=	°F + 459 58
°C	=	5/9 (°F − 32)
°K	=	5/9°R

Table 13. Length and Area

1 statute mile (mi)	=	5280 feet
	=	1.609 kilometers
1 foot (ft)	=	12 inches
	=	30.48 centimeters
1 inch (in)	=	25.40 millimeters
100 ft per min	=	0.508 meter per sec
1 square foot	=	144 sq inches
	=	0.0929 sq meter
1 square inch	=	6.45 sq centimeters
1 kilometer (km)	=	1000 meters
	=	0.621 statute mile
1 meter (m)	=	100 centimeters (cm)
	=	1000 millimeters (mm)
	=	1.094 yards
	=	3.281 feet
	=	39.37 inches
1 micron	=	0.001 millimeter
	=	0.000039 inch
1 meter per sec	=	196.9 ft per min

Table 14. Horsepower Equivalent

1 HP	=	33,000 ft. lb. per min.
1 HP	=	550 ft. lb. per sec.
1 HP	=	2,546 Btu per hr.
1 HP	=	42.4 Btu per min.
1 HP	=	.71 Btu per sec.
1 HP	=	746 watts

Table 15. Approximate Metric Equivalents

1 Decimetre	=	4 inches
1 Metre	=	1.1 yards
1 Hectare	=	2-1/2 acres
1 Litre	=	1.06 qt.
1 Kilogramme	=	2.2 lb.
1 Metric Ton	=	2,200 lb.

Table 16. Miscellaneous Measures

Angles or Arcs

60 seconds (") =	1 minute
60 minutes (') =	1 degree
90 degrees (°) =	1 rt. angle or quadrant
360 degrees =	1 circle

Avoirdupois Weight

437.5 GRAINS (gr.) =	1 ounce
16 ounces (7,000 grains) =	1 pound
2,000 pounds =	1 short ton
2,240 pounds =	1 long ton

Cubic Measure

2.728 cubic inches (cu. in.) =	1 cubic foot
27 cubic feet =	1 cubic yard

Square Measure

144 square inches (sq. in.) =	1 square foot
9 square feet =	1 square yard

ELECTRIC

Table 17. Appliance Energy Requirements

Major Appliances	Annual kWh
Air-Conditioner (room) (Based on 1000 hours of operation per year. This figure will vary widely depending on geographic area and specific size of unit)	860
Clothes Dryer	993
Dishwasher (including energy used to heat water)	2,100
Dishwasher only	363
Freezer (16 cu. ft.)	1,190
Freezer-frostless (16.5 cu. ft.).	1,190
Range with oven	700
with self-cleaning oven	730

Table 17. Appliance Energy Requirements (*Continued*)

Major Appliances	Annual kWh
Refrigerator (12 cu. ft.)	728
Refrigerator-frostless (12 cu. ft.)	1,217
Refrigerator/Freezer (12.5 cu. ft.)	1,500
Refrigerator/Freezer-frostless (17.5 cu. ft)	2,250
Washing Machine-automatic (including energy used to heat water)	2,500
Washing Machine-non-automatic (including energy to heat water)	2,497
washing machine only	76
Water Heater	4,811

Kitchen Appliances

Blender	15
Broiler	100
Carving Knife	8
Coffee Maker	140
Deep Fryer	83
Egg Cooker	14
Frying Pan	186
Hot Plate	90
Mixer	13
Oven, Microwave (only)	190
Roaster	205
Sandwich Grill	33
Toaster	39
Trash Compactor	50
Waffle Iron	22
Waste Disposer	30

Heating and Cooling

Air Cleaner	216
Electric Blanket	147
Dehumidifier	377
Fan (attic).	281
Fan (circulating)	43
Fan (rollaway)	138

(*Continued*)

Table 17. Appliance Energy Requirements (*Continued*)

Major Appliances	Annual kWh
Fan (window)	170
Heater (portable)	176
Humidifier	163

Laundry

Iron (hand)	144

Health and Beauty

Germicidal Lamp	141
Hair Dryer	14
Heat Lamp (infrared)	13
Shaver	1.8
Sun Lamp	16
Toothbrush	.5
Vibrator	2

Home Entertainment

Radio	86
Television	
Black and White	
Tube type	350
Solid State	120
Color	
Tube type	660
Solid State	440

Housewares

Clock	17
Floor Polisher	15
Sewing Machine	11
Vacuum Cleaner	46

Table 18. Alternating Current Calculations

Current To Calculate	Alternating Three-Phase	Single Phase
Amperes when horsepower is known	$\dfrac{\text{H.P.} \times 746}{1.73 \times E \times \%\text{Eff} \times \text{P.F.}}$	$E \times \%\text{Eff} \times \text{P.F.}$
Amperes when kilowatts are known	$\dfrac{kW \times 1000}{1.73 \times E \times \text{P.F.}}$	
Amperes when K.V.A. are known	$\dfrac{\text{K.V.A.} \times 1000}{1.73 \times E}$	
Kilowatts	$\dfrac{I \times E \times 1.73 \times \text{P.F.}}{1000}$	
K.V.A	$\dfrac{I \times E \times 1.73}{1000}$	
Horsepower (Output)	$\dfrac{I \times E \times 1.73 \times \%\text{Eff} \times \text{P.F.}}{746}$	

E = Volts.
kW = Kilowatts.
P.F. = Power Factor.
I = Amperes

%Eff. = Percent Efficiency.
K.V.A = Kilovolt amperes.
H.P. = Horsepower.

Table 19. Kilowatt Conversion Factors

Kilowatt Conversion Factors

1 kilowatt	=	1.3415 horsepower
	=	738 ft lb per sec
	=	44,268 ft lb per min
	=	2,656,100 ft lb per hr
	=	56.9 Btu per min
	=	3,413 Btu per hr

Table 20. Effect of Electrical Current on Humans

Current Values	Effect
1 ma	Causes no sensation
1 to 8 ma	Sensation of shock. Not painful
8 to 15 ma	Painful shock
15 to 20 ma	Cannot let go
20 to 50 ma	Severe muscular contractions
100 to 200 ma	Ventricular fibrillation
200 & over ma	Severe burns. Severe muscular contractions

Table 21. Equivalent Electrical Units

1 Kilowatt	= 1,000 Watts
1 Kilowatt	= 1.34 H.P.
1 Kilowatt	= 44,260 Foot-Pounds per minute
1 Kilowatt	= 56.89 B.T.U. per minute
1 H.P.	= 746 Watts
1 H.P.	= 33,000 Foot-Pounds per minute
1 H.P.	= 42.41 B.T.U. per minute
1 Btu	= 778 Foot-Pounds
1 Btu	= 0.2930 Watt-Hour
1 Joule	= 1 Watt-Second

Table 22. United States Power Characteristics

	Voltage	Amperes	Phase
Controls	20 to 120	5 to 15	Single
Small Equipment	120 208 240 277	10 to 40	Single
Large Equipment	208 240 480	30 to 400	Three

Table 23. Induction Motor Synchronous Speeds

POLES	@ 60 Hz
2	3,600
4	1,800
6	1,200
8	900
10	720
12	600

FORMULA

Table 24. Geometric Formulas

Circumference of a circle	$C = \pi d$
Length of an arc	$L = \dfrac{n}{360} \times \pi d$
Area of a rectangle	$A = LW$
Area of a square	$A = s^2$
Area of a triangle	$A = 1/2 bh$
Area of a trapezoid	$A = 1/2 h(b + b')$
Area of a circle	$A = .7854 d^2$, or $1/4 \pi d^2$
Area of a sector	$S = \dfrac{n}{360} \times .7854 d^2$
Area of an ellipse	$A = .7854 ab$
Area of the surface of a rectangular solid	$S = 2LW + 2LH + 2WH$
Lateral area of a cylinder	$S = \pi dh$
Area of the surface of a sphere	$S = \pi d^2$
Volume of a rectangular solid	$V = LWH$
Volume of a Cylinder	$V = .7854 d^2 h$
Volume of a sphere	$V = .5236 d^3$, or $1/6 \pi d^3$
Volume of a cube	$V = e^3$

HEAT

Table 25. Heat Equivalents

1 Btu	=	252 calories
1 kilocalorie	=	1000 calories
1 Btu/lb.	=	.55 kcal/kg
1 Btu/lb.	=	2.326 kj/kg
1 kcal/kg	=	1.8 Btu/lb
1 Btu/hr	=	0.2931 watts

Table 26. Heat Generated By Appliances

General lights and heating	3.4 Btu/hr/watt
2650 watt toaster	9100 Btu/hr
5000 watt toaster	19,000 Btu/hr
Hair dryer	2000 Btu/hr
Motor less than 2 HP	3600 Btu/hr/HP
Motor over 3 HP	3000 Btu/hr/HP

Table 27. Heat Loss From Hot Water Piping

Pipe Size Inches	Hot Water, 180°F Bare	Insulated
1/2	65	22
3/4	75	25
1	95	28
1-1/4	115	33
1-1/2	130	36
2	160	42
2-1/2	185	48
3	220	53
4	280	68

Table 28. Heat Loss From Bare Steel Piping

(Btu/hr @ 70 deg F ambient)

	Hot Water		Steam	
Nominal Pipe Size	120 F	180 F	5 psi	100 psi
1/2	0.455	0.546	0.612	0.760
1	0.684	0.819	0.919	1.147
2	1.180	1.412	1.578	1.987
4	2.118	2.534	2.850	3.590
8	3.880	4.638	5.210	6.610
12	5.590	6.670	7.500	9.530

Table 29. Specific Heats of Common Substances

Aluminum	.2143
Brine	.9400
Coal	.314
Copper	.0951
Ice	.5040
Petroleum	.5110
Water	1.0000
Wood	.3270

Table 30. Heat Content of Common Fuels

Number 6 fuel oil	152,400 Btu per gallon
Number 2 fuel oil	139,600 Btu per gallon
Natural Gas	950 to 1150 Btu per cubic foot
Propane	91,500 Btu per gallon

Table 31. Typical Fuel Oil Analysis

	California Hi Sulphur/Lo Sulphur		Texas Hi Sulphur/Lo Sulphur	
% Sulphur	1.0	/4.2	1.0	/2.8
% Total Ash	0.1	/0.08	0.08	/0.10
% Carbon	87.2	/85.2	86.8	/86.1
% Hydrogen	10.0	/10.0	10.8	/10.1
% Oxygen/Nitrogen	1.8	/1.0	1.0	/1.0
Moisture	0.3	/		/0.2
Specific Gravity (@60F)	1.007	/0.986	0.977	/1.003

MISCELLANEOUS

Table 32. Cylindrical Tank Capacity

Capacity (gal)	Diameter (in)	Length
100	24	4'
150	30	4'
250	30	7'
500	42	7'
550	48	6'
750	48	8'
1000	65	6'
1500	48	16
2000	65	11' 11.5"
3000	65	17' 10"
4000	65	23' 8.5"
5000	72	23' 9"
10000	108	21'

Table 33. Metric Capacity

Name	Capacity
Milliliter (ml.)	= .001
Centiliter (cl.)	= .01
Liter (1)	= 1.
Decaliter (Dl.)	= 10
Hectoliter (Hl,)	= 100
Kiloliter (Kl.)	= 1,000
Myrialiter (Ml.)	= 10,000

Table 34. Metric Length

	Meters
Millimeter (mm.)	= .001
Centimeter (cm.)	= .01
Decimeter (dm.)	= .1
Meter (m)	= 1
Decameter (Dm)	= 10
Hectometer (Hm.)	= 100
Kilometer (Km.)	= 1,000
Myriameter (Mm.)	= 10,000

Table 35. Metric Weight

Name	Grams
Milligram (mg.)	.001
Centigram (cg.)	.01
Decigram (dg.)	.1
Gram (g)	1
Decagram (Dg.)	10
Hectogram (Hg.)	100
Kilogram (Kg.)	1,000
Myriagram (Mg.)	10,000
Quintal (Q.)	100,000
Tonneau (T.)	1,000,000

Table 36. Weights of Metals

Name of Metal	Pounds/Cu. Ft.
Aluminum	166
Brass	504
Copper	550
Iron	450
Lead	712
Silver	655
Steel	490
Tin	458
Zinc	437

Table 37. Ultimate Bending Strength

Material	PSI
Cast iron	30,000
Wrought iron	45,000
Steel	65,000
Stone	1,200
Concrete	700
Ash	8,000
Hemlock	3,500
Oak, white	6,000
Pine, white	4,000
Pine, yellow	7,000
Spruce	3,000
Chestnut	4,500

Table 38. Recommended Lighting Levels

Area	Foot-Candles
Perimeter of building	5
Office areas	70
Corridors, elevators and stairways	20
Toilets and washrooms	30
Entrance lobbies	10
Dining areas	20
Mechanical rooms	20

Appendix F—List of Tables

STEAM

Table 39. Steam Pressure Temperature Relationship

Gage PSI	Sat Temp F	Gage PSI	Sat Temp F
0	212	70	316
5	228	80	324
10	240	90	331
20	259	100	338
30	274	200	388
40	287	300	422
50	298	400	448
60	308		

Table 40. Steam Trap Selection

Characteristic	Inverted Bucket	Thermostatic
Method of Operation	Intermittent	Intermittent
Steam Loss	None	None
Resistance to Wear	Excellent	Good
Corrosion Resistance	Excellent	Excellent
Resistance to Hydraulic Shock	Excellent	Poor
Vents Air and CO_2 at Steam Temperature	Yes	No
Ability to Vent Air at Very Low Pressure	Poor	Excellent
Ability to Handle Start-up Air Loads	Fair	Excellent
Operation Against Back Pressure	Excellent	Excellent
Resistance to Damage from Freezing	Poor	Excellent
Ability to Purge System	Excellent	Good
Ability to Operate on Very Light Loads	Good	Excellent
Responsiveness to Slugs of Condensate	Immediate	Delayed
Ability to Handle Dirt	Excellent	Fair
Comparative Physical Size	Large	Small

TEMPERATURE

Table 41. Color Scale of Temperature

Color	Temperature
Incipient red heat	900-1100
Dark red heat	1100-1500
Bright red heat	1500-1800
Yellowish red heat	1800-2200
Incipient white heat	2200-2600
White heat	2600-2900

Table 42. Centigrade/Fahrenheit Scale

°C	°F
–50	–58
–40	–40
–30	–22
–20	–4
–10	14
0	32
10	50
20	68
30	86
40	104
50	122
60	140
70	158
80	176
90	194
100	212
110	230
120	248
130	266
140	284
150	302
160	320

Table 43. Melting Point of Common Substances

Metal	Symbol	Degrees F
Aluminum	Al	1218
Copper	Cu	1981
Iron	Fe	2795
Lead	Pb	621
Mercury	Hg	-38
Molybdenum	Mo	4750
Silicon	Si	2590
Silver	Ag	1761
Tin	Sn	449
Tungsten	W	6100
Zinc	Zn	787

Table 44. Temperature of Waste Heat Gases

Source of Gas	Temperature, Deg. F
Ammonia oxidation process	1,350 - 1,475
Annealing furnace	1,100 - 2,000
Black liquor recovery furnace	1,800 - 2,000
Cement kiln (dry process)	1,150 - 1,350
Cement kiln (wet process)	800 - 1,100
Coke oven	
beehive	1,950 - 2,300
by-product	up to 750
Copper refining furnace	2,700 - 2,800
Copper reverberatory furnace	2,000 - 2,500
Diesel engine exhaust	550 - 1,200
Forge and billet heating furnace	1,700 - 2,200
Garbage incinerator	1,550 - 2,000
Gas benches	1,050 - 1,150
Glass tanks	800 - 1,000
Heating furnace	1,700 - 1,900
Malleable iron air furnace	2,600
Nickel refining furnace	2,500 - 3,000
Petroleum refinery still	1,000 - 1,100
Steel furnace, open hearth	
oil, tar, or natural gas	800 - 1,100
producer gas-fired	1,200 - 1,300
Sulfur, ore processing	1,600 - 1,900
Zinc fuming furnace	1,800 - 2,000

WATER

Table 45. Water Equivalents

U. S. Gallons	× 8.33 = Pounds
U. S. Gallons	× 0.13368 = Cu. Ft.
U. S. Gallons	× 231. = Cu. Ins.
U. S. Gallons	× 3.78 = Liters
Cu. Ins. of Water (39.2°)	× 0.03613 = Pounds
Cu. Ins. of Water (39.2°)	× 0.004329 = U. S.Gals.
Cu. Ins. of Water (39.2°)	× 0.576384 = Ounces
Cu. Ft. of Water (39.2°)	× 62.427 = Pounds
Cu. Ft. of Water (39.2°)	× 7.48 = U. S. Gals.
Cu. Ft. of Water (39.2°)	× 0.028 = Tons
Pounds of Water	× 27.72 = Cu. Ins.
Pounds of Water	× 0.01602 = Cu. Ft.
Pounds of Water	× 0.12 = U. S. Gals.

Table 46. Supply Line Sizes For Common Fixtures

Laundry Tubs	1/2 inch
Drinking Fountains	3/8 inch
Showers	1/2 inch
Water-Closet Tanks	3/8 inch
Water-Closets (with flush valves)	1 inch
Kitchen Sinks	1/2 inch
Commercial-Type Restaurant Scullery Sinks	1/2 inch

Table 47. Pressure of Water

One foot of water = 0.4335 psi   One psi = 2.31 foot of water

Feet Head	Pressure PSI
10	4.33
15	6.49
20	8.66
25	10.82
30	12.99

(Continued)

Table 47. Pressure of Water (*Continued*)

Feet Head	Pressure PSI
35	15.16
40	17.32
45	19.49
50	21.65
55	23.82
60	25.99
70	30.32
80	34.65
90	38.98
100	43.31
200	86.63
300	129.95
400	173.27

Table 48. Water Requirements of Common Fixtures

Fixture	Cold, GPM	Hot, GPM
Water-closet flush valve	45	0
Water-closet flush tank	10	0
Urinals, flush valve	30	0
Urinals, flush tank	10	0
Lavatories	3	3
Shower, 4-in. head	3	3
Shower, 6-in. head and larger	6	6
Baths, tub	5	5
Kitchen sink	4	4
Pantry sink	2	2
Slop Sinks	6	6

Table 49. Blowdown Flow Rates-3" Pipe

PSI	GPS	PSI	GPS
15	0.50	60	1.06
20	0.60	70	1.13
30	0.73	80	1.20
40	0.86	90	1.30
50	0.96	100	1.36

Table 50. Tank Capacities Per Foot of Depth

Diameter in Feet	Gallons
1	5.84
2	23.43
3	52.75
4	93.80
5	146.80
6	211.00
7	287.00
8	376.00
9	475.00
10	587.00
12	845.00
14	1,150.00
16	1,502.00
18	1,905.00
20	2,343.00

Table 51. Efficiency Loss Due To Scale

Thickness in Inches	Percent Loss
1/64	4
1/16	11
1/8	18
3/16	27
1/4	38
3/8	48
1/2	60

GLOSSARY

A.B.M.A.—American Boiler Manufactures Association

ABSOLUTE HUMIDITY—the weight in grains of water vapor actually contained in 1 cubic foot of the air and moisture mixture.

ABSOLUTE PRESSURE—atmospheric pressure added to gage pressure

ABSOLUTE TEMPERATURE—the theoretical temperature when all molecular motion of a substance stops. Minus 460 degrees Fahrenheit

ACCESS FLOORING—a raised floor consisting of removable panels under which ductwork, wiring and pipe runs are installed

ACID CLEANING—a process in which dilute acid, used in tandem with a corrosion inhibitor, is applied to metal surfaces for removing foreign substances too firmly attached.

ACUATOR—A device that converts a pneumatic or electric signal to force which produces movement

A.E.E.—Association of Energy Engineers

AHERA—Asbestos Hazard Emergency Response Act (1986); federal law requiring local education agencies to identify asbestos hazards and develop abatement plans.

AHU—air handling unit

AIR CHANGES—the number of times in an hour that a volume of air filling a room is exchanged

AIR GAP—the radial space between the rotating element and the stationary element of a generator or motor, through which space the magnetic energy passes.

ALGAE—a form of plant life which causes fouling in water system piping; especially in cooling towers

ALKALINE—a condition of liquid, opposite from acidic on the PH scale, which is represented by carbonates, bicarbonates, phosphates, silicates or hydroxides contained within it.

AMBIENT TEMPERATURE—the temperature of the air immediately surrounding a device

AMPLIFIER—a device that receives an input signal from an independent source and delivers an output signal that is related to, and generally greater than the input signal.

ANALOG—a continuous range of values such as temperature, pressure, etc. (contrast with binary).

ANEMOMETER—an instrument used for measuring air velocity

ANTHRACITE COAL—a dense coal known for its low volatility which enables the use of smaller combustion chambers than those needed to burn bituminous coal.

APPARENT POWER—a term used to describe the product of current and voltage expressed in KVA

APRON—A paved area, such as the juncture of a driveway with the street or with a garage entrance.

ATMOSPHERIC PRESSURE—the weight of the atmosphere measured in pounds per square inch

ATOMIZATION—the process of reducing a liquid into a fine spray

AUTO-TRANSFORMER—a transformer of single coil construction in which both primary and secondary connections are made to the same coil at different taps

AUTOMATIC VARIABLE-PITCH FAN—a propeller type fan whose hub incorporates a mechanism which enables the fan blades to be re-pitched simultaneously and automatically

AUXILIARY DEVICE—A component, added to a control system, which when actuated by the output signal from one or more controllers produces a desired function.

AWG—American wire gauge

AXIAL FAN—a device which discharges air parallel to the axis of its wheel

BACKING PLATE—a steel plate positioned behind a welding groove to confine the weldment and assure full penetration

BACKWASH—the backflow of water through the resin bed of a water softener during the cleaning process

BAFFLE—a structure or partition used for directing the flow of gasses or liquids

BAGASSE—the dry pulp remaining from sugar cane after the Juice has been extracted; a fuel used in boiler furnaces

BALANCED DRAFT—a fixed ratio of incoming air to outgoing products of combustion

BATT—insulation in the form of a blanket, rather than loose filling

BDC—bottom dead center; when a piston is at the bottom of its stroke

BEARING WALL—a wall structure which supports floors and roofs

BHP—brake horsepower; the actual power produced by an engine

BIOCIDE—a substance that is destructive to living organisms that is used in refrigeration systems by design

BITUMINOUS COAL—a soft coal generally more volatile and requiring larger combustion chambers in which to burn than anthracite coal

BLEED-OFF—water discharged from the system to control concentrations of salts or other impurities in the circulating water

BLISTER—a raised area on the surface of metal caused by overheating

BLOWBACK—the difference in pressure between when a safety valve opens and closes

BLOWDOWN—the removal of water from a boiler in lowering its chemical concentrations

BOILER HORSEPOWER—the evaporation of 34.5 pounds of water per hour from a temperature of 212°F into dry saturated steam

BOILING OUT—a process whereby an alkaline solution is boiled within a vessel to rid its interior of oil or grease

BOILING POINT—The temperature at which a liquid is converted to a vapor corresponding to its pressure

BOYLES LAW—a law of physics dealing with variations in gas volumes and pressures at constant temperatures

BRAKE HORSEPOWER—the actual power output of a motor, turbine or engine

BREECHING—a large duct used for conveying gasses of combustion from a furnace to a stack

BRITISH THERMAL UNIT (Btu)—a unit measurement of heat. The amount of heat needed to raise the temperature of one pound of water, one degree Fahrenheit

BUNKER COIL—residual fuel oil of high viscosity commonly used in marine and stationary steam power plants. (#6 fuel oil.)

BUS—vertical and horizontal metal bars which distribute line-side electrical power to branch circuits

BUSHING—a removable sleeve inserted or screwed into an opening to limit its size

BUTTERFLY VALVE—a throttling valve consisting of a centrally hinged plate that can be opened partially or expose the full cross section of the pipe it feeds by maneuvering the valve handle

through a quarter turn.
BX—electrical cable wrapped in rubber with a flexible steel outer covering
CALORIE—the quantity of heat needed to raise the temperature of one gram of water, one degree centigrade
CANTILEVER—A projecting beam or Joist, not supported at one end, used to support an extension of a structure.
CAPACITOR—a device capable of storing electric energy consisting of two conducting surfaces separated by an insulating material. It blocks the flow of direct current while allowing alternating current to pass.
CAPILLARY ACTION—the capacity of a liquid to be drawn into small spaces
CARRYOVER—a condition whereby water or chemical solids enter the discharge line of a steam boiler
CASING—the outer skin or enclosure forming the outside of an appliance
CATALYST—any substance, usually a solid, used to accelerate or retard a chemical reaction which does not itself undergo change during the process.
CAVITATION—the formation of vapor pockets in a flowing liquid
CELL—smallest tower subdivision which can function as an independent unit with regard to air and water flow.
CERCLA—Comprehensive Environmental Responsibility, Compensation and Liability Act (1976); also Superfund: federal law authorizing identification and remediation of abandoned hazardous waste sites.
CFC—chlorofluorocarbon; chemical substance associated with depletion of Earth's ozone layer.
CFM—cubic feet per minute
CH RATIO—carbon-hydrogen ratio
CHASSIS—the frame or plate on which the components of a device are mounted
CHIMNEY EFFECT—the tendency of air to rise within confined vertical passages when heated
CHLORINATION—the addition of the chemical chlorine to water
C I D—cubic inch displacement
CLOSED LOOP SYSTEM (HVAC)—The arrangement of components to allow system feedback, e.g., a heating unit, valve and thermostat arranged so that each component affects the other and can

react to it.

COAGULATION—the initial aggregation of finely suspended matter by the addition of floc forming chemical or biological process

COAXIAL CABLE—Cable that consists of a tubular conductor surrounding a central conductor held in place by insulating material. Used for transmitting high frequency signals.

COEFFICIENT OF HEAT TRANSMISSION (U)—the amount of heat measured in Btu's transmitted through materials over time. The heat transmitted in one hour per square foot per degree difference between the air inside and outside of a wall, floor or ceiling.

COEFFICIENT OF PERFORMANCE (COP)—the ratio of work performed to the energy used in performing it

COMBUSTION EFFICIENCY—the ratio of the heat released from a fuel as it burns, to its heat content

COMPOSITE FUEL—fuel resulting from blending finely ground coal with oil—a coal-oil mix; other combinations of fuel such as a blend of coal and municipal waste

CONCENTRATIONS—the number of times that dissolved solids increase in a body of water as a one to one multiple of the original amount due to the evaporation process

CONDENSATION—the process of returning a vapor back to its liquid state through the extraction of latent heat

CONDUIT—a pipe, tube or tray in which electrical wires are run and protected

CONTINUOUS BLOWDOWN—a process whereby solids concentrations within a body of water are controlled through the constant removal and replacement of the water

CONVECTION—a process of heat transfer resulting from movement within fluids due to the relative density of its warmer and cooler parts

CORROSION—the wasting away of metals due to physical contact with oxygen, carbon-dioxide or acid

COUNTERFLOW—a method of heat exchange that brings the coldest portion of one moving fluid into contact with the warmest portion of another

CPU—central processing unit. That portion of a computer which contains the arithmetic and logic functions which process programmed instructions

CRUDE OIL—unrefined petroleum in its natural state as it comes from the ground

CURING COMPOUND-a liquid which is sprayed onto new concrete to prevent premature dehydration

DAMPER—a mechanism used to create a variable resistance within a gas or air passage in order to regulate its rate of flow

DB—decibel unit for describing noise level

DEAERATION—the removal of entrained air from a liquid

DEDICATED—Set apart or committed to a definite use (wiring, conduit, etc.). Addressed to a definite task (signals).

DEGREE DAY—a unit representing one degree of difference from a standard temperature in the average temperature of one day, used to determine fuel requirements

DEGREE OF SUPERHEAT—the difference between the saturation temperature of a vapor and its actual temperature at a given pressure

DELTA CONNECTION—a three-phase connection in which the start of each phase is connected to the end of the next phase forming the Greek letter. The load lines are connected to the corners of the delta.

DEMINERALIZATION—deionization. The removal of ionizable salts from solution

DESICCANT—a drying agent such as silica gel or activated alumina that is used to absorb and hold moisture

DEW POINT TEMPERATURE—the lowest temperature that air can be without its water vapor condensing

DIELECTRIC STRENGTH—the ability of insulation to withstand voltage without rupturing.

DISSOLVED SOLIDS—solids such as sodium chloride, calcium bicarbonate, etc., which are dissolved in water similar to the dissolving of sugar in water.

DIRECT CURRENT—an electrical current which flows in only one direction

DISTRIBUTION BASIN—shallow pan-type elevated basin used to distribute hot water over the tower fill by means of orifices in the basin floor.

DOT—U.S. Department of Transportation; enforces regulations governing the transport of hazardous materials.

DPDT SWITCH—double pole double throw switch

DRE—destruction and removal efficiency; measure of the effectiveness of incineration in removing contaminants from waste materials.

DRIFT—circulating water lost from the tower as liquid droplets en-

Glossary

trained in the exhaust air steam

DRIFT ELIMINATORS—which the air passes prior to its exit from the tower, for the purpose of removing entrained water droplets from the exhaust air

DRY BULB TEMPERATURE—the temperature of the air as measured on a thermometer

DUCT FURNACE—a furnace located in the ducting of an air distribution system to supply warm air for heating

DVW—drain, vent, waste pipes

ECONOMIZER—a heat recovery device that utilizes waste heat for preheating fluids

EFFICIENCY—heat output per hat input. For boilers, Btu in steam output per hour divided by Btu in fuel input per hour

EFFLUENT—Treated sewage from a septic tank or sewage treatment plant.

ELECTRIC IGNITION—ignition of a pilot or burner by an electric spark generated by a transformer

ELECTROLYSIS—a chemical reaction between two substances prompted by the flow of electricity at their point of contact

ELECTROSTATIC PRECIPITATOR—an electrically charged device used for removing dust particles from an air stream

EMF—electromotive force. Voltage

ENTHALPY—The actual or total heat contained in a substance. It is actually calculated from a base. In refrigeration work the base temperature is accepted as -40 (degrees F) -40 (degrees C).

ENTHALPY SWITCHOVER—Automatic switching or regulation of outside air and return air dampers. Total heat content of inside and outside air is compared before selecting either inside or outside or a mixture of the two for ventilating, which will require the least amount of refrigeration, humidification or dehumidification.

ENTRAINMENT—the inclusion of water or solids in steam, usually due to the violent action of the boiling process

EPA—U.S. Environmental Protection Agency; primary federal agency responsible for enforcement of federal laws protecting the environment; also referred to as the Agency.

ETHANOL—The two-carbon atom alcohol present in the greatest proportion upon the fermentation of grain and other renewable resources such as potatoes, sugar or timer. It is sometimes called grain alcohol.

EVACUATION—The removal of air and moisture from a refrigeration system.

EVAPORATION—the transformation of a liquid into its vapor state through the application of latent heat

EXCESS AIR—the air supplied for the combustion process in excess of that theocratically needed for complete oxidation

FACE AND BYPASS—Duct and damper system which directs air through (face) or around (bypass) heating or cooling coils.

FAIL SAFE CONTROL—A device that opens an electric circuit when the sensing element senses an abnormal condition.

FARAD—The unit of electrical capacity of a capacitor.

FAN PITCH—the angle which the blades of a propeller fan make with the plane of rotation, measured at a prescribed point on each blade

FEEDWATER TREATMENT—the conditioning of water with chemicals to establish wanted characteristics

FIFRA—Federal Insecticide, Fungicide and Rodenticide Act (1972, 1988); federal law mandating toxicity testing and registration of pesticides.

FILL—that portion of a cooling tower which constitutes its primary heat transfer surface. Sometimes referred to as "packing".

FIRE SEPARATION WALL—a wall dividing two sections of a building used to prevent the spread of fire

FLAME ROD—a metal or ceramic rod extending into a flame which functions as an electrode in a flame detection circuit

FLAME SAFEGUARD SYSTEM-the equipment and circuitry used to provide safe control of burner operation

FLAME SIMULATOR—a device used as a substitute for the presence of flame to test a flame detection circuit

FLASHBACK—the backward movement of a flame through a burner nozzle

FLASH POINT—the minimum temperature necessary for a volatile vapor to momentarily ignite

FLOATING ACTION—Movement of the controlled device either toward its open or its closed position until the controller is satisfied, or until the controlled device reaches the end of its travel or until a corrective movement in the opposite direction is required. Generally there is a neutral zone in which no motion of the controlled device is required by the controller.

FLOCCULANT—a chemical used to bridge together previously coagulated particles

FLOW RATE—the amount of fluid passing a given point during a specified period to time

FLUE—a pipe or conduit used for conveying combustion exhaust fumes to the atmosphere

FLUME—a trough which may be either totally enclosed, or open at the top used in cooling towers for primary supply of water to various sections of the distribution system

FLUX DENSITY—magnetic lines of force per unit of area

FOAMING—the continuous formation of bubbles having a high surface tension which are hard to disengage from a surface

FORCED DRAFT—the process of moving air mechanically by pushing or drawing it through a combustion chamber with fans or blowers

FOULING—the accumulation of refuse in gas passages or on heat absorbing surfaces which results in undesirable restrictions to flow

FUSIBLE PLUG—A safety plug used in refrigerant containers that melts at a high temperature to prevent excessive pressure from bursting the container.

GAGE PRESSURE—absolute pressure minus atmospheric pressure

GAUGE MANIFOLD—A manifold that holds both the pressure and compound gauges, the valves that control the flow of liquids.

GPG—grains per gallon. One grain per gallon equals 17.1 ppm

GPM—gallons per minute

GASAHOL—blends of alcohols and gasoline to produce fuel for internal combustion engines

GRAIN—A unit of weight equal to 1/7000 lb (0.06480 g), which is used to indicate the amount of moisture in the air.

GRAVITY FEED—the transfer of a liquid from a source to an outlet using only the force of gravity to induce flow

GROOVING—a form of corrosion wherein a groove is formed along the length of tubes or shells

GROUND—an electrical connection made between any structure or object and the earth

HALOGENS—chlorine, iodine, bromine or fluorine

HALIDE TORCH—a device which uses an open flame for detecting refrigerant leaks

HARDNESS—a term used to describe the calcium and magnesium content of water

HEADER—a manifold to which many branch lines are connected

HEAT EXCHANGER—a device used to transfer heat from one medium to another

HEAT GAIN, TOTAL—the sum total of sensible plus latent heat gain from ventilation and infiltration of outside air (convection), heat conduction through walls and roofs, solar radiation and heat generated by people, lights and machinery.

HEATING SURFACE—that portion of a heat exchange device which is exposed to the heat source and transfers heat to the heated medium

HERMETIC COMPRESSOR—a unit wherein a compressor and its driving motor are contained in a single, sealed housing

HERTZ—(Hz) one cycle per second Hg (mercury)—A silver-white, heavy, liquid metal. It is the only metal that is a liquid at room temperature.

HIGH FIRE—the firing rate at which a burner consumes the most fuel thus producing the most heat

HIGH LIMIT—the maximum value at which a controller is set that if exceeded causes the shut down of a system

HIGH SIDE—The part of the refrigeration system that contains the high pressure refrigerant. Also refers to the outdoor unit, which consists of the motor, compressor, outdoor coil, and receiver mounted on one base.

HIGH-SIDE CHARGING—The process of introducing liquid refrigerant into the high side of a refrigeration system. The acceptable manner for placing the refrigerant into the system.

HIGH TEMPERATURE BOILER—a boiler which produces hot water at pressures exceeding 160 psi or at temperatures exceeding 250 degrees Fahrenheit

HORSEPOWER—(hp)—a unit of power equal to 550 foot pounds per second, 33,000 foot pounds per minute or 746 watts.

HRS—Hazard Ranking System; system used to rank NPL sites in terms of degree of contamination and urgency for remediation.

HRT—horizontal return tubular boiler

HSWA—Hazardous and Solid Waste Amendments; 1984 amendments to RCRA establishing a timetable for landfill bans and more stringent requirements for USTs.

HUMIDISTAT—a control device which responds to changes in the humidity of air

HWTC—Hazardous Waste Treatment Council; Washington-based trade association of more than 60 treatment and disposal firms.

HYDROSTATIC TEST—a procedure in which water is used to determine the integrity of pressure vessels

IC—internal combustion

ID—inside diameter

IGNITION TEMPERATURE—the minimum temperature at which the burning process can begin for a given fuel source

INDUCED DRAFT FAN—a fan or blower located in the breeching of gas passages that produces a negative pressure in the combustion chamber causing air to be drawn through it

INERTS—non-combustible particulates found in fuel

INTERLOCK—a sensor or switch which monitors the status of a required condition which causes a programmed action to occur when the condition becomes inappropriate

INTERRUPTING CAPACITY—A rating given to a piece of electrical equipment. The rating represents the maximum short circuit conditions under which the equipment will not disintegrate. This rating must be equal to or greater than the short circuit current available at the particular building location.

I/O DEVICES—a device used to convey information to or from a computer e.g. a keyboard or printer. Input/output

ION EXCHANGE—a process for removing impurities from water on the atomic level through the selective repositioning of electrons.

IONIZATION DETECTOR—Detects fire in its very beginning stages, before smoke, flame or appreciable heat is present by detecting invisible products of combustion (referred to as the incipient stage).

INTERFACE—A shared boundary. An interface might be a hardware component to link two devices or it might be a portion of storage or registers accessed by two or more computer programs.

JUMPER—a short length of wire used to by-pass all or part of an electrical circuit

KELVIN SCALE—a temperature scale incremented in centigrade that begins at absolute zero (–273C)

KNOCKOUT—portal designed into the side of an electrical box or metal cabinet that can be easily removed to accommodate wires or piping KVA—kilovolt amperes

LATENT HEAT OF CONDENSATION—the heat extracted from a vapor in changing it to a liquid with no change in temperature

LATENT HEAT OF EVAPORATION—the heat added to a liquid in changing it to a vapor with no change in temperature

LATENT HEAT OF FUSION—the heat added to a solid in changing it to a liquid with no change in temperature

LIGNITE—coals characterized by high moisture content and/or high volatiles. Lignite may be defined as a solid fuel more mature than peat but less mature than bituminous coal.

LITHIUM BROMIDE—a chemical having a high affinity for water used as a catalyst in absorption refrigeration systems

LOAD SHEDDING—Shutting down non-critical electrical equipment to a preselected level, when a peak electric use period approaches. This is done to prevent paying an excessive electric rate which is based on highest electric usage during a billing period.

LOCKED ROTOR—a test in which a motors rotor is locked in place and rated voltage is applied

LOCKED ROTOR AMPS—the amperage which is apparent in a live circuit of a motor driven device when the rotor is not moving

LOWER—An opening with horizontal slats to permit passage of air, but excluding rain, sunlight and view.

LOW LIMIT—the minimum value at which a controller is set that if dropped below will result in a shut down of the system

LOW PRESSURE BOILER—a steam boiler whose maximum allowable working pressure does not exceed 15 psi.

LOW SIDE—Those parts of a refrigeration system in which the refrigerant pressure corresponds to the evaporating coil pressure.

LOW-SIDE CHARGING—The process of introducing refrigerants into the low side of the system. Usually reserved for the addition of a small amount of refrigerant after repairs.

LOW VOLTAGE RELAY—(Integrated Control Center)—A relay which initiates load shutdown whenever line voltage is less than 83% of the normal line voltage.

LOW WATER CUTOFF—a mechanism used for shutting off the supply of fuel to a furnace when a boiler's water level falls to a dangerously low level

LPG—liquified petroleum gas

MAKEUP WATER—water added to a system to replace that which was lost due to leaks, consumption, blow down and evaporation MANOMETER—a U-shaped tube used for measuring pressure differences in air passages

MANUAL RESET—the operation required after a system undergoes a safety shutdown before it can be put back into service

MANUFACTURED GAS—fuel gas manufactured from coal, oil, etc., as differentiated from natural gas

MASTER—An instrument whose variable output is used to change the set point of a submaster controller. The master may be a humidistat, pressure controller, manual switch, transmitter, thermostat, etc.
MAWP—maximum allowable working pressure
MEASURED VARIABLE—The uncontrolled variable such as temperature, relative humidity or pressure measured by a sensing element.
MEGA—one million times
MEGOMETER—an instrument used for evaluating the resistance values of electrical wire coverings
METHANOL—A one-carbon atom alcohol formerly prepared by the destructive distillation of wood. Formerly called wood alcohol.
MHO—a unit measurement of electrical conductance
MICRON—one millionth of a meter. 1/25,400 in.
MIXING VALVE—a three-way valve having two inlets and one outlet designed specifically for mixing fluids
MODULATING FIRE—varying the firing rate with the load thereby decreasing the on-off cycling of burners
N.E.C.—National Electrical Code
NATURAL CIRCULATION—the circulation of fluids resulting from differences in their density NC—normally closed. A relay contact which is closed when the relay coil is not energized
NCP—National Contingency Plan (National Oil and Hazardous Substances Pollution Contingency Plan); regulations promulgated by EPA to implement CERCLA and Sec. 311 of CWA.
NITROGEN BLANKET—a technique used whereby the air space above a body of water in a vessel is filled with nitrogen to keep oxygen from coming into contact with its metal surfaces
NO—normally open. A relay contact which is open when the relay coil is not energized
NOMINAL DIMENSION—an approximate dimension. A conventional size
NONCONDENSABLE GAS—Any gas, usually in a refrigeration system, that cannot be condensed at the temperature and pressure at which the refrigerant will condense, and therefore requires a higher head pressure.
NORMALLY CLOSED—Applies to a controlled device that closes when the signal applied to it is removed. (closing a pneumatic device stops the flow of the control-agent-closing an electrical

device allows an electrical current to flow)

NORMALLY OPEN—Applies to a controlled device that opens when the signal applied to it is removed. (opening a pneumatic device allows the flow of the control agent)-opening an electrical device stops the flow of electrical current)

NPDES—National Pollutant Discharge Elimination System; federal permitting system required by EPA for hazardous effluents.

NPL—National Priorities List; official list of hazardous waste sites to be addressed by CERCLA.

NPT—national pipe thread

NRC—Nuclear Regulatory Commission; federal body overseeing the operation of nuclear power plants and other installations.

OD—outside diameter.

OHM—a unit measurement of electrical resistance

OHM'S LAW—A mathematical relationship between voltage, current, and resistance in an electrical circuit. It is stated as follows: voltage (E) = amperes (I) × ohms (R).

OIL BINDING—A condition in which a layer of oil on top of liquid refrigerant may prevent it from evaporating at its normal pressure and temperature.

ONE PIPE SYSTEM—a system in which one pipe serves as both the supply and return main

OPEN CIRCUIT—an electrical circuit in which the current path has been interrupted or broken

ORSAT—a device used to analyze gasses by absorption into chemical solutions

OSHA—Occupational Safety and Health Administration; oversees and regulates workplace health and safety.

OVERLOAD ALARM CIRCUIT—(Integrated Control Center)—Used to sound an alarm, light a pilot light locally or send a communique to a central building automation console. Is usually mechanically linked to the overload contacts of the starter in an integrated control center.

OVERLOAD PROTECTOR—a safety device designed to stop motors when overload conditions exist.

OVERLOAD RELAY—a relay which operates to interrupt excessive currents

OVERSHOOT—The greatest amount a controlled variable deviates from its desired value before stabilizing, after a change of input.

OXYGEN SCAVENGER—a chemical treatment such as sulfite or hydrazine used for releasing dissolved oxygen from water

PACKAGE BOILER—one that is shipped from the assembly plant completely equipped with all the apparatus needed for its operation

PANEL HEATING—a method whereby interior spaces are heated by pipe coils located within walls, floors and ceilings

PCB—polychlorinated biphenyl; a pathogenic and teratogenic industrial compound used as a heat-transfer agent; PCBs may accumulate in human or animal tissue.

PE—pneumatic/electric relay

PEAK LOAD—the maximum load carried for a stated short period of time

PERFECT COMBUSTION—the complete oxidation of a fuel using no excess air in the combustion process

PERIPHERAL DEVICE—a hardware item forming part of a computer system that is not directly connected to but supports the processor

PF—power factor

PH—a value that indicates the intensity of the alkalinity or acidity of a solution

PHASE ANGLE—the relation between two sinusoidally varying quantities of the same frequency which do not pass through zero at the same instant. One cycle is considered to contain 360 electrical degrees. The extent by which the zero points differ is expressed as a part of the total of 360 degrees

PHASE ROTATION—the sequence in which the phases of a generator or network pass through zero points of their waves. The same sequence must exist on all units which are to be paralleled.

PILOT—a small burner used as an igniter to light off a main burner

PILOT TUBE—A device that measures air velocity.

PLENUM CHAMBER—a compartment to which ducts are connected enabling the distribution of air to more than one area

POINTING—Treatment of joints in masonry by filling with mortar to improve appearance or protect against weather.

POLY-PHASE MOTOR—an electric motor driven by currents out of phase from circuit to circuit.

POSITIVE DISPLACEMENT—an action wherein the total amount of a fluid being transferred by a mechanical device is accomplished without leakage or back siphonage

POTENTIOMETER—a variable resistor in an electrical circuit

POWER FACTOR—an efficiency value assigned to electrical circuits based on a comparison of its true and apparent power characteristics

PPM—parts per million

PRECIPITATION—the removal of constituents from water by chemical means. Condensation of water vapor from clouds

PRESSURE, LIMITER—A device that remains closed until a predetermined pressure is reached and then opens to release fluid to another part of the system or opens an electrical circuit.

PRESSURE REGULATOR—a mechanism used to maintain a constant pressure within a feeder line regardless of fluctuations above the setting in the supply line

PRIMARY AIR—combustion air which is introduced into a furnace with the fuel

PRIMING—the discharge of water particles into a steam line

PRODUCER GAS—An industrial fuel made by processing air and steam continuously through the hot fuel bed of a gas producer. Consists essentially of carbon monoxide and hydrogen (50 percent) and nitrogen (50 percent)

PRODUCTS OF COMBUSTION—any gas or solid remaining after the burning of a fuel

PROXIMATE ANALYSIS—test for fuels whose results are related to the performance of a fuel upon heating and burning. Properties analyzed are moisture, volatile matter, ash and fixed carbon. Usually used in connection with ultimate analysis to aid in the evaluation of coals.

PRV—pressure reducing valve

PSI—pounds per square inch

PSYCHROMETRIC CHART—a graph which depicts the relationship between the pressure, temperature and moisture content of air

PSYCHROMETER—an instrument incorporating both a dry-bulb and a wet-bulb thermometer, by which simultaneous dry-bulb and wet-bulb temperature readings can be taken

PURGE—eliminating a fluid from a pipe or chamber by flushing it out with another fluid

PVC—polyvinyl chloride

RACEWAY—Any method commonly used and accepted for running electrical wires or cable or pneumatic lines within a building. Raceway may be exposed or concealed in floors, walls, ceiling plenums, or may be buried or installed in outdoor locations.

RADIATION LOSS—the loss of heat from an object to the surrounding air

RATED CAPACITY—the manufacturers stated capacity rating for mechanical equipment, for instance, the maximum continuous capacity in pounds of steam per hour for which a boiler is designed.

RAW WATER—water supplied to the plant before any treatment

RCRA—Resource Conservation and Recovery Act (1980); regulates management and disposal of hazardous materials and wastes currently being generated, treated, disposed or distributed.

RECTIFICATION—the conversion of alternating current to direct Current

REED VALUES—A piece of thin, flat, tempered steel plate fastened to the valve plate.

REFRACTORY—heat resistant material used to line furnaces, ovens and incinerators

REFRIGERATION—the removal of heat from an area where it is not wanted to one that is not objectionable

REGISTER—the grill work or damper through which air is introduced

RELAY—an electromechanical device having a coil which, when energized and de-energized, opens and closes sets of electrical contacts

RELIEF VALVE—a device used to relieve excess pressure from liquid filled pressure vessels, pipes and hot water boilers

RETORT—any closed vessel or facility for heating a material for purposes of chemical reaction.

REVERSE ACTING—the output signal changing in the opposite direction the controlled or measured variable changes. Example: an increase in the controlled or measured variable results in a decreased output signal.

RH—relative humidity

RINGELMANN CHART—a comparator of smoke density comprised of rectangular grids filled with black lines of various widths on a white background

RMS—root mean square

RPM—revolutions per minute

RUNNING CURRENT—the amperage draw noted when a motor is running at its rated speed

SAFETY FACTOR—the ratio of extra strength or capacity to the calculated requirements to ensure freedom from breakdown and ample capacity.

SAFETY PLUG—a device that releases the contents of a container to prevent rupturing when unsafe pressures or temperatures exist.

SAFETY VALVE—a quick-opening safety valve used for the fast relief of excessive pressure in a container.

SAIL SWITCH—a switch attached to a sail shaped paddle inserted into an air stream which is activated when the air stream striking the sail reaches a pre-established velocity

SARA—superfund Amendments and Reauthorization Act (1986); federal law reauthorizing and expanding the jurisdiction of CERCLA.

SARA Title III—part of SARA mandating public disclosure of chemical information and development of emergency response plans.

SATURATED STEAM—dry steam which has reached the temperature corresponding to its pressure

SBI—Steel Boiler Institute

SCALE—a hard coating or layer of chemical materials on internal surfaces of pressure vessels, piping and fluid passages

SCHRADER VALVE—a spring-loaded valve that permits fluid to flow in only one direction when the center pin is depressed.

SCOTCH MARINE BOILER—a horizontal fire tube boiler of the wet wall type

SEDIMENTATION—settling out of particles from suspension in water

SENSIBLE HEAT—heat which changes the temperature but not the state of a substance

SENSIBLE HEAT RATIO—the percentage of total heat removed that is sensible heat. It is usually expressed as a decimal and is the quotient of sensible heat removed divided by total heat removed.

SET POINT—a pre-determined value to which a device is adjusted that, when reached, causes it to perform its intended function

SHORT CIRCUIT—an unintentional connection between two points in an electrical circuit resulting in an abnormal flow of current

SHUNT TRIP—a coil mechanism which trips the breaker in response to an external voltage potential being applied to the coil.

SINGLE PHASE—A circuit or a device energized by a single alternating voltage. One phase of a polyphase system

SLAVE—an instrument using the signal of a master unit to expand the master unit capabilities.

SLING PSYCHROMETER—a device having a wet and a dry bulb thermometer which measures relative humidity when moved rapidly through the air

SLUGGING—a condition in which a quantity of liquid enters the compressor cylinder, causing a hammering noise.

SNG—synthetic natural gas

SPALLING—deterioration of materials evidenced by flaking of their surfaces

SPECIFIC GRAVITY—the ratio of the weight of any substance to the same volume of a standard substance at the same temperature

SPECIFIC HEAT—a measure of the heat required in Btu's to raise the temperature of one pound of a substance, one degree Fahrenheit. The specific heat of water is 1.0

SPEED REDUCER—a mechanical device, incorporated between the driver and the fan of a mechanical draft tower, designed to reduce the speed of the driver to an optimum speed for the fan

SPLIT-PHASE MOTOR—a motor with two windings. Both windings are used in the starting of the motor. One is disconnected by a centrifugal switch after a given speed is reached by the motor. The motor then operates on only one winding.

SSU—Seconds Saybolt Universal

STANDARD ATMOSPHERE—a condition existing when the air is at 14.7 psia of pressure, 68 (degrees F) 20 (degrees C) temperature, and 36% relative humidity.

STATIC HEAD—the pressure exerted by the weight of a fluid in a vertical column

STATIC PRESSURE—the force exerted per unit area by a gas or liquid measured at right angles to the direction of flow

STEAM SEPARATOR—a device for removing the entrained water from steam

STRATIFICATION OF AIR—a condition of the air when little or no movement is evident

SUMP—a container, compartment or reservoir used as a drain or receptacle for fluids

SUPERHEAT—a temperature increase above the saturation temperature or above the boiling point.

SURGE SUPPRESSOR—a device capable of conducting high transient voltages, which protects the other devices that could be destroyed thereby

SYNTHETIC FUELS—fuels which exist in physical and chemical forms different from those in the naturally occurring state. Any fuel made from coal, unconventional oil resources or fermented farm products

TDC—top dead center. When a piston is at the top of its stroke

TDS—total dissolved solids

TENSILE STRENGTH—the capacity of a material to withstand being stretched

TERTIARY AIR—air supplied to a combustion chamber to supplement primary and secondary air

THERM—a symbol used in the industry representing 100,000 Btu's.

THERMISTOR—a solid state device whose electrical resistance varies with temperature

THERMOCOUPLE—a mechanism comprised of two electrical conductors made of different metals which are joined at a point which, when heated, produces an electrical voltage having the value that is directly proportional to the temperature of the heat being applied

THERMOPILE—a battery of thermocouples connected in series

THERMOSTATIC EXPANSION VALVE—a control device operated by the pressure and temperature of an evaporator which meters the flow of refrigerant to its coil

THREE-MODE SWITCHING—three-state switching operations such as fast-slow-stop, summer-winter-auto, day-night-auto, etc.

TON OF REFRIGERATION—the heat required to melt a one ton block of ice in 24 hours. 288,000 Btu's or 12,000 Btu's per hour

TOTAL SOLIDS CONCENTRATION—the weight of dissolved and suspended impurities in a unit weight of boiler water, usually expressed in parts per million

TRANSDUCER—an instrument that converts an input signal into An output signal, usually of another form. e.g. electrical input to pneumatic output.

TRANSIENT—a temporary change in steady-state conditions occurring during load changes

TWO POSITION VALVE—a valve which is either fully open or fully closed, having no positions in between

UNINTERRUPTIBLE POWER SUPPLY—a separate source of electricity used to maintain continuity of electrical power to a device or system when the normal supply is interrupted

UNITARY SYSTEM—a combination heating and cooling system that is factory assembled in one package and is usually designed for conditioning one room or space.

ULTIMATE ANALYSIS—a chemical analysis of a fuel with respect to carbon, hydrogen, oxygen, nitrogen, sulphur and ash

VACUUM—any pressure less than that of the surrounding atmosphere

VAPOR LOCK—the vaporizing of pumped liquid causing a shut down or slowing of the pumping process

VAPOR RETARDER—a barrier constructed of materials which retard the capillary action of water into building structures

VAR—volt-ampere reactive

VAV—variable air volume

VELOCIMETER—an instrument used to measure the speed of moving air

VENTURI—a short tube designed with a constricted throat that increases the velocity of fluids passed through it.

VISCOSITY—a measure of a fluids resistance to flow

VOLTAGE DIP—the instantaneous reduction in voltage resulting from an increase in load before the regulator corrects it

VOLTAGE DROOP—the difference in voltage from no-load to full-load. May be expressed in percentage by dividing by the full load voltage and multiplying by 100 percent

VOLTAGE DROP—the reduction in voltage caused by a flow of current through a resistance in DC circuits or impedance in AC circuits. Equal to the product of current and resistance or impedance

WATER WALL—that portion of the steam generator water tube that forms the wall of the combustion chamber

WET BULB TEMPERATURE—the lowest temperature that can be attained by an object that is wet and exposed to moving air

WHEATSTONE BRIDGE—a laboratory instrument used for measuring resistance

WIRING HARNESS—a pre-assembled group of wires of the correct length and arrangement to facilitate interconnections

WORK RELAY—a device using a current sensing principle to detect the level of work a particular motor is doing. Control contacts of the device can activate a local indicator or can be tied into a central building automation system's console. The device is also equipped with an analog output.

WORKING RANGE—the desired controlled or measured variable values over which a system operates.

WYE CONNECTION—same as star connection. A method of interconnecting the phase of a three-phase system to form a configuration resembling the letter Y. A fourth or neutral wire can be connected to the center point

INDEX

A
A/C unit 213, 214
absorption 101
access panels 114
access doors 95
accessories 113
accident investigation 141
aftercoolers 114
air compressor 201, 202
air balancing 126
air handling units 8, 125, 126, 202
air dryers 8
air handlers 8
antifreeze protection 110
apprenticeship 85
asbestos management 10
ASHRAE 104
ASME 97, 149
automatic expansion valve 104
auxiliaries 73, 74, 101, 113, 114
auxiliary 113, 114
auxiliary units 2, 4, 5, 8, 9, 28

B
battery 118, 129, 245, 256
bearings 115, 120, 122, 124, 253, 254
bells and whistles 125
bid package 55
blow by 132
blowdown valves 149
boiler room 73, 89, 101
boiler explosion 149
boiler plate 52, 175

boiler control failures 146
boilers 89-100, 136, 158, 155, 168, 202
boiling point 131
B & M insurance 139, 141
building engineers 73
building envelope 165
burner control 148

C
calibration 31, 129, 132, 158, 256-258
call back 33
capillary tubes 104
causes of accidents 143
caveat emptor 58
centrifugal force 101
centrifugal chillers 5
centrifugal pumps 119
chemical storage 11
chemical treatment 4
chemicals 91
chief engineer 71-74, 136, 137
chilled water system 109
chiller 101, 135, 155, 203, 204
circulating pumps 135
co-generation 118
code compliance 33
cold weather operation 109
combustion air 89, 93
combustion theory 225
combustion 223-226
combustion efficiency 100
complete combustion 225

compression systems 101
compressor 113, 156
computerized maintenance 53
concept maintenance 158
condensate return 4, 9, 149
condensate system 89
condensers 102, 122
conservation measures 170, 172
contingency planning 11, 24, 153
contract scope 56
contracts 21, 27, 42
controls 122
cooling system 119
cooling tower 9, 204, 205
cooling season 114
corrosion 219-223
CPU 58

D
data entry 62
definition of accident 140
degree of superheat 104
department operations 18, 20, 80
design pressure 95
dewpoint temperature 108
direct expansion 102
domestic water 9
draft system 91
dry bulb temperature 108
dye penetrant 160

E
eddy current 160, 161
electric motor 115-117, 205, 206
electric power 126, 156
electric panel 206, 207
electrical system protection 127, 128
electrodes 148
electromechanical rooms 125, 136

electromechanical failures 145
elevators 9
emergency power 7
emergency generator 73
emergency instructions 154-157
emergency procedures 81
employee productivity 57
employee orientation 22
energy consumption 165
energy management 18, 25, 174
equipment rooms 133
equipment operation 10, 71, 90
equipment shut downs 76
equipment operation 71, 90
equipment conservation 151
equipment entries 60
equipment operation 113
equipment management 22
evaporators 102
exhaust fans 9, 125
exits 98

F
facilities management 66, 71, 79, 82, 104
Facilities Manager 1, 19, 31
fan speed 110
fan imbalances 28, 110
feed water regulator 91, 94
feedwater 90
field inspection 50
fire protection 18, 24
fire alarm system 7
fireside 92
firetube boiler 93
firing rates 95, 96
flame safeguard 28
flame impingement 92
flexible couplings 120
floor drains 9

forecasted budget 55
foundation 123
free cooling 109
fresh air intakes 9
fuel oil transfer 4
fuel system 89, 91, 147
furnace explosions 149
fusible plug 130

G
gage glass 97, 149
gear oils 228
generator 209, 210
generator maintenance 118

H
hardware requirements 58, 59
heat transfer efficiency 100
heat exchanger 89, 210
hermetic 101
high pressure side 104
high water 97
horsepower 116
hot water by-pass 110
humidity ratio 108
HVAC 169, 173, 174

I
impellers 101, 120
implementation plan 12, 13
impulse trap 132
in-service training 22
incinerator 9, 211
incoming power 7
incomplete combustion 226
infrared scan 128
inspection 128, 247-253
installation specs 105
instrumentation 95, 153
insulation 128

insurance 35, 44, 94, 106
insurance carrier 91
intercoolers 114
internal examination 94
inverters 128

J
job instructions 60, 65
job descriptions 21, 64, 66-74
journal 1
journeymen mechanics 64

L
laboratory analysis 31
ladders and walkways 98, 99, 110
licenses and permits 75, 76, 83, 106
load shedding 169
log entries 107
loss report 141-143
loss prevention 139
low water 97, 146
low pressure side 104
low water cut off 91, 94, 147, 148
lube oil system 123
lubricant storage 228
lubricants 226
lubrication 159, 226-229
lubricators 114

M
magnetic particle 161
main distribution panel 7
maintenance 128, 166
maintenance management 56
maintenance procedures 54
make-up 4, 28
manpower requirements 63, 64
master equipment list 58, 60
materials management 57
mechanical refrigeration 104, 107
mechanical compression 101

mechanical trap 132
mechanical seals 120
mechanics of burning 223
metal fatigue 115
metering devices 104
MG set 128
minimum clearance 99
motor generator 128
motors 246
MSDS 81

N
natural gas 157
nondestructive testing 159
normal operating level 147
nuisance tripping 10, 153

O
oil seals 124
oil storage tanks 95
operating engineers 72, 76-78
operating inspection 94
operating speed 58
operating efficiency 99
operating certificate 75, 76
operating schedule 78
operating manuals 32
operations and maintenance 87
organizational requirements 20
orientation and training 79, 82-87

P
packing 120
peak shaving 169
perfect combustion 225
performance bond 36
performance appraisals 65
peripheral 58, 60
personal protective gear 137
plant management 79, 80
plant engineering 71

plumbing system 156
PM procedure 118
PM mechanic 78
PM function 53
policies and procedures 15-19, 104, 152, 158
policy development 19
policy topics 20-25
policy manuals 80
pop action 130
position descriptions 65
positive displacement 101
power house 1, 2, 10, 27, 72, 73, 80, 152
power plant 54, 65, 75, 89, 101
power measurement 128
pressure settings 132
pressure 114
pressure vessel 144, 145
pressure variations 135
pressure-relief valves 130-132
preventive maintenance 23, 54, 58, 63, 74, 77, 107, 126, 201
productivity 55
programs 61
properties of oil 226
proposals 28-32, 51, 61
protective devices 153, 159
PRV 109
psychrometry 108
pulse-echo 162
pump repairs 119-121
pump packing 11
pumps 214, 215
purging 95

Q
quality assurance 18, 25
quick drain test 148, 149
quotations 28, 32-34

Index

R
radiography 161, 162
record keeping 152
reference library 71
refractory 96
refrigerant 108
refrigeration system 101
refrigeration 215
regulatory compliance 12, 25
relative humidity 108
relief coverage 77
relief valve 130
repair histories 31
reporting systems 11
RFP 56
rotary system 128
roundsman 136
rupture disk 130

S
safety devices 123, 140
safety controls 130, 131
safety valve 130, 136, 150, 257
safety/control system 91
safety management 18, 23
safety valve testing 11
scale formation 150
scale 219
scheduled shut downs 12
scope of work 35
service agreements 18, 25
service schools 83
shell and tube 102
shift engineer 63
shift firemen 74
shut downs 152
site inspection 41
slow drain test 149
software 54-61
soot blowers 97

SOP manual 15, 83
specifications 33-51, 56-58
standard operating procedures 15
stationary engineer 77, 231-244
steam generators 95, 101
steam trapping 132
steam boilers 3
steam system 100
steam distribution 4, 90, 91
steam traps 91
structural integrity 9, 114
support equipment 113
switching mechanisms 159
system modules 57
system components 101
system types 125
system commissioning 107
system design 152
system checks 245

T
temperature 114
thermocycle cooling 109
thermographic studies 28, 128, 162
thermostatic trap 132
thermostatic expansion valve 104
thickeners 226
training program 83
training 56
transfer pump 136
transfer switch 247
troubleshooting 152
turbine operation 122-124
turbine inspection 121, 122

U
ultrasonics 162, 163
UPS systems 128
utility interruptions 24

V

vacuum 133, 140, 156
valve and switch lists 22, 135
valves 9
vendor lists 11
vendor qualifications 29
vibration analysis 163, 164
vibration 135, 136, 153

W

watch engineer 135

water-steam-water cycle 90
water level 96, 135
water treatment 12, 28, 217, 219
waterside 94, 100
ways of travel 98
weak link failures 152
wearing rings 120
wet bulb temperature 108
work control 57
work backlogs 55
work order system 22